Design for Environmental Sustainability

Carlo Vezzoli • Ezio Manzini

Design for Environmental Sustainability

Carlo Vezzoli
Ezio Manzini

Design and Innovation for Sustainability
Politecnico di Milano
INDACO Department
20158 Milan
Italy

ISBN 978-1-84800-162-6 e-ISBN 978-1-84800-163-3

DOI 10.1007/978-1-84800-163-3

British Library Cataloguing in Publication Data
Vezzoli, Carlo
Design for environmental sustainability
1. Sustainable design 2. Design, Industrial - Environmental aspects
I. Title II. Manzini, Ezio
745.2
ISBN-13: 9781848001626

Library of Congress Control Number: 2008930165

Authorized translation from Italian language edition published by Zanichelli editore SpA, Bologna, Italy (*Design per la Sostenibilita Ambientale,* 2007, ISBN 978-88-08-16744-6).

Translation: Kristjan Pruul

Cover image: The MULO system, a four-wheeled hybrid convertible vehicle powered by solar, electric and human power. The MULO system was designed by Fabrizio Ceschin of the Design and System Innovation for Sustainability research unit of the Politecnico di Milano Department of Industrial Design (INDACO), in collaboration with the "A. Ferrari" istituto Professionale Statale per l'Industria e l'Artigianato (IPSIA), Maranello.

Image research: Sara Cortesi, Lucia Orbetegli, Carlo Proserpio

Cover design: eStudio Calamar S.L., Girona, Spain

Printed on acid-free paper

9 8 7 6 5 4 3 2 1

springer.com

To my daughters Giada and Viola

United Nations Decade of Education for Sustainable Development

The goal of the United Nations Decade of Education for Sustainable Development (2005–2014, DESD), for which UNESCO is the lead agency, is to integrate the principles, values, and practices of sustainable development into all aspects of education and learning.

This educational effort will encourage changes in behaviour that will create a more sustainable future in terms of environmental integrity, economic viability, and a just society for present and future generations.

www.unesco.org/education/desd/

Preface and Acknowledgements

This book was born from close co-operation between two authors. However, the first part is written mainly by Ezio Manzini, meanwhile the second, third, fourth and appendixes are written by Carlo Vezzoli.

The book is based on the work, research and didactic activities of the Research Unit *Design and Innovation for Sustainability* (DIS), and the *Environmental Requirements for Industrial Products* Laboratory (RAPI.labo) of INDACO Department at the *Politecnico di Milano University* and the Italian National Network of Universities for Design for Sustainability (RAPI.rete). A valuable contribution comes from the collaboration of many years within these structures with Lucia Orbetegli and Carlo Proserpio.

The first part outlines and defines the main features and scenarios of sustainable development; proposes new ideas for well-being in sustainable society; describes the roles of various stakeholders inside system innovation, paying particular attention to the function of the designer.

The second part looks into the approach to and various strategies for designing environmentally sustainable products. With special emphasis on the Life Cycle Design approach, the design criteria and guidelines for integrating environmental requirements as early as the development stage will be explained. System design for eco-efficiency is also introduced.

The third part presents several methods and tools for the development and assessment of low impact products, services and systems, stressing the Life Cycle Assessment method.

In the fourth part the historical course of research, education and practice of design for sustainability is outlined.

The appendixes summarise the design strategies/criteria and guidelines, and outline the environmental effects.

This text was published and written in order to offer both a comprehensive and organic picture of the topic of design for environmental sustainability as well as a supporting handbook in design. For this reason, as well as the general discussion, possible strategies, guidelines and design options are also listed. For the very same reason, the overall text is illustrated with a healthy amount of examples of low impact products. These examples are chosen for a dual purpose: to clarify design options and their position with regard to various other types of products. To make things easier to understand in a visual way, and to provide access on different levels, these examples are listed against a coloured background.

Some arguments that are closely related to the subject, but more detailed than others, are also presented in deliberately distinctively coloured boxes.

The reading of this book can be greatly assisted by employing the auxiliary design software, which is freely available at www.lens.polimi.it. Especially worth mentioning here are:

- ICS_Toolkit – qualitative tool for creating environmentally sustainable product concepts (coherent with criteria and guidelines of Chaps. 5–9).
- SDO_MEPSS – qualitative tool for creating environmentally sustainable system concepts (as approached in Chap. 10).

Introduction

The combination of environment and design has been termed *ecodesign*. Even if, as it appears, it would be more precise to use other, more specific terms, let us start with few ideas on *ecodesign* as it allows us to define better our application field (and because *ecodesign* is still widely used). So what is *ecodesign*?

It is unquestionable that on the first comprehensive level the word *eco-design* is rather self-explanatory, as its general meaning derives directly from its component terms: *eco-design* is design following ecological criteria. Manifested as a compound expression of a vast complex of design activities, it tends to handle ecological questions with an upstream approach, *i. e.* redesigning the products themselves.

But at the same time, *eco-design* is one of those terms that, even if it represents certain ideas, it is far away from giving any precise definition. In this case, its insecure state is born from carrying and amplifying the vast semantic fields of ecology and design that in turn are not free from indetermination either. It appears that, when the expression has entered into research and professional circles and finally into institutional documents, then whoever speaks using these terms refers to theoretical hinterlands and introverted fields that very often vary from speaker to speaker.

If *eco-design* means a general designing tendency, conscious of the environmental impacts of its products, we can easily judge that it represents a rather blurry concept (and behind it rather large set of indications).

It is well known that *design* in its full meaning concerns any planning activity ranging from urban and territorial design to visual or architectural design up to the designing of consumable goods.

On the other hand, extension of environmental issues that cross all topics that arise actually allows to the terms of ecology and design to be bound in all those diverse surroundings of design activities for as long as possible, with an accurate analysis of specific practices, identifying a block of questions associated or associable with the environment.

Besides, it is not only possible, but already a reality, in a sense, that there is no domain of design where, at least on a debatable level, the question about relationships between a specific sector and environmental issues has not been raised.

But here the range of attention will be restricted to the complex design activity that mainly deals with manufacturing industrial goods, or rather the field that is currently referred to as *industrial design.*

This limitation of observation platform, even while excluding other important sectors of design, maintains a remarkable internal articulation and complexity, due to either a wide range of interested professionals or a degree of operational field or cultural background from which it originates.

Not only this though. The *industrial design* will be intended in its most up-to-date definition, which does not apply merely to physical products (defined by material, shape and function), but which also extends to the *production system,* that is, the integrated body of goods, services and communication that is used to represent the companies on the market.

Environmental awareness and derived activities have followed an upstream route: from pollution treatments (*end-of-pipe* policies rely on downstream neutralisation of negative environmental effects created by industrial products) to intervention in production processes that cause pollution (the topic of *clean technologies),* to redesigning the products and/or services that make these processes necessary (the topic of *low impact products*). Finally, the ecological awareness has brought to discussion and to reorientation social behaviour, *i. e.* the demand for the products and services that ultimately motivate the existence of those processes and products (the topic of *sustainable consumption*).

Consequently, this progression has involved a certain transformation of the overall nature of the participating variables: when two primary levels (downstream interventions and *clean technologies*) focus mainly on technological issues, then in successive ones (*low impact products* and *sustainable consumption*) the role of design and socio-cultural issues amplifies progressively. As a matter of fact, low impact product development might also require *clean technologies*, but demands for secure new designing capacities (in fact, it is possible to create *low impact products* without any technological sophistication). Similarly, even with more emphasis, promoting sustainable consumption and behaviour might create a demand for new products, but it can also entail reorientation of choices towards new product service systems (that together satisfy certain needs and desires) that in order to be accepted require cultural and behavioural changes of the consumers. Thus, suggesting on these grounds that solutions that introduce higher ecological qualities cannot disregard the level of socio-cultural susceptibility.

Within this general frame of reference the role of industrial design can be summarised as activity that connects *technologically possible* with *ecologically necessary* and tends to give birth to new significant socio-cultural propositions.

To understand better what has been said, it is useful to indicate and briefly present four fundamental levels of intervention:

- Environmental *redesign* of existing systems (choosing low impact materials and energy)
- Designing new products and services (substituting old systems with more environmentally sustainable ones)
- Designing new production–consumption systems (offering *intrinsically sustainable* satisfaction of needs and desires)
- Creating new scenarios for sustainable life style.

Environmental redesign of existing systems: considering the life cycle of the examined product, it attempts to improve global efficiency with the selection of low impact materials and energy sources.

The first stage thus entails mainly technological characteristics (*e. g.* atoxical, renewable, biodegradable, recyclable resources) and does not require actual changes in consumer behaviour or life style. Here, references to social components or the market are limited to a common ecological sensibility spread in the demand–supply of environmentally sustainable goods, choosing between otherwise similar products (mainly the cases in which *environmental labels* are put to use).

Its limits lie in the need to position the low impact product solutions in the system that is developed and founded by sources without any environmental concern.

More of the propositions active today (or already in the elaborative stage) act on following level: from those promoted by the main producers (as in the field of home appliances or cars, just to name a few more important cases) through those developed by research centres, to those promoted by single designers or small- and medium-scale companies motivated by good-will.

Designing new products and services: considering a given demand for efficiency, it attempts to develop new products and services that could environmentally succeed the existing ones. It entails the adoption of an approach and tools orientated to regard every possible environmental implication connected with every stage of a project's life cycle (pre-production, production, distribution, use and disposal) until the final phase of the project, and to minimise negative effects. For this approach the term *Life Cycle Design* has been coined.

The second level intervention requires the new propositions to be recognised as valid and socially acceptable.

Operating on this level of technical-productive innovation can be more easily directed towards environmentally qualitative research, if existing products can no longer be redesigned. Although it does not question the result-orientated demand it refers to, it is still necessary to expect difficulties introducing environmentally sound products and services into a cultural and behavioural context that is based on different expectations and values.

Significantly few concrete initiatives are active on this level, though some cases of great interest can be seen on the horizon.

Designing new product service systems: considering the demand for *satisfaction* as variable, it attempts to offer different (and more sustainable) ways of obtaining results that could become socially appreciated and at the same time radically favourable for the environment.

The third intervention level requires the new blends of proposed goods and services (a new system of products and services) to become so socially appreciated that they can overcome the cultural and behavioural inertia of the consumers. In order to be effective such a design range has to be positioned on a *strategic* level of the system: the designer or the company (or association of companies) set out for such promotion has to accept the investment risks of a product without any verified market, knowing that, if successful, the possibilities of opening new markets might arise.

Thus, the designing and estimating of these products and service systems is based on strategic dimensions, but the life cycle approach can also be employed, *i. e. Life Cycle Design*, for the block of products and supporting products for services that make up such a supply system.

Creating new scenarios for a "sustainable life style": this attempts to develop activities in a cultural sphere that could promote new qualitative criteria and in perspective modify the same result-orientated demand–supply structure.

The fourth intervention level cannot emerge without the complex dynamics of socio-cultural innovation, which could provide the inner role (small, albeit important) for designers to collect, interpret, recalculate and stimulate socially created ideas. In this case, the focus is not as much on the introduction of recent technological or production solutions, but on promoting new qualitative criteria that at the same time are environmentally sustainable, socially acceptable and culturally attractive.

On this level, design activities can occur in various forms while being only indirectly related to production (*e. g.* articles, books, conferences, exhibitions). It is a level where design research organisations play the key roles, but relationships can also be created with companies that would like to redefine their identity and in perspective also their cultural role.

Today the expression *design for environmental sustainability* is used to refer to this block of activities and to those various levels of intervention that have the aforementioned variable dimension of changes required by the transition towards sustainable society.

Thus, an intention to *design for environmental sustainability* entails facilitating the capacity of the production system to respond to the social demand of well-being, while using drastically smaller amounts of environmental resources than needed by the present system. This requires coordinated management of all available tools and provides homogeneity and visibility for all proposed products, services and communication.

This is the heart of the book: a contribution to the design culture and practice, capable of handling the transition to sustainability and to promote the new, emerging generation of intrinsically sustainable goods and services.

Contents

Part I
Frame of Reference

Chapter 1
Sustainability and Discontinuity

1.1 Introduction

For the last 50 years, humankind has faced possible self-destruction and found itself in an unforeseen condition of being conscious about it, *i. e.* knowing that nuclear weapons and the ecological crisis can not only change the path of history, but also become the end of history (as in "the day after" there will be no us – humans, to tell the story).

It is still too early to judge how much this discovery has changed our culture or how much it will still change. But it is certain that this new condition is not going to pass: once our capacity to hurt ourselves has reached this point, it is no longer possible to return. From here onwards, among the possible scenarios open to humankind, will be one without future.

This confirmation certainly has a tragic dimension, but is not necessarily dominated/ruled by pessimism. The ability of humans to hurt themselves is a characteristic rooted in their ways of being: Taking this into account is not a pessimist presumption, on the contrary, it is more realistic to note an inevitable component of human nature. From the other side, human nature is also capable of loving, feeling and, put prosaically, uses its intelligence to search for a way to hurt itself as little as possible. Also, it can and has to be aware of this. The lives of individuals and social groups are always immersed in the plotting and combination of these two components. And this is how it is always going to be. A new question arises about how they are going to combine with each other in the face of the understanding that the first can destroy the world, in case it surpasses the second.

Nobody can foresee this. Only to have faith: the appearance of possible self-destruction in future scenarios might help to grow collective wisdom. Just as understanding eventual death can help an individual live his or her life more wisely (and appreciate it more); thus, acknowledging the possibility of collective death might promote a society that would be "wiser" and more capable of understanding the value of those unstable conditions called "peace" and "social and ecological equity".

There is little sense in discussing the actual probability of such a course of action. It is pointless because, in the light of implications brought by other scenarios, it is evident that it is the only direction in which to head. Accepting this point of view will create a new question: how can we favour the positive side of human nature? Or not to sound boastful, how can we promote such cultures and organisations that would hinder people from doing too much harm to themselves?

Before answering those questions it would be useful to remember that nuclear destruction and ecological catastrophes have threatened us in diverse times and due to diverse reasons. Now from a certain distance, we realise that they are two aspects of the same phenomenon and with an ability to listen, we would understand that they teach us the same thing: both are unwanted creatures of human intelligence, that show us the power and limits at the same time (limits of rationality and capacity to control our choices in the medium to long term). Both play their part in a certain social and productive system and in its way of existing and functioning. Certainly, one appears in a scenario of violent war and another in the peaceful search for well-being. But a well-focused observation would reveal immediately how this search for well-being has intertwined with violence, how arms make part of our productive system (one of the most profitable parts), and how their utilisation is part of economical and commercial politics. Both tell us that something has to be changed.

From this meditation many different discourses may start. Ours will have its primary stress on ecological topics and on sustainable changes, and will try to shed light on their connection with other topics as well; this is more than anything the goal of this book, i. e. guidelines for the design of sustainable goods and systems.

1.2 Sustainable Development and Environmental Sustainability

In 1987 the World Commission for Environment and Development (WCED) prepared a document entitled *Our Common Future*, also known as the *Brundtland Report*, after the coordinator of the commission, Gro Harlem Brundtland. It was the first occasion on which the concept of *sustainable development* was introduced.

Its definition says in a few words the following: *sustainable development is development that meets the needs of the present without compromising the ability of future generations to meet their own needs* (WCED, 1987).

Evidently this definition does not speak about the environment *per se,* but refers to the well-being of people as an environmental quality. From this originates a fundamental ethical principle: the responsibility of present generations to future generations.

Our Common Future is an historically relevant document for two reasons. First, it brought into international debate the aforementioned definitive imposition of the idea about responsibility for the future. Second, which in the end might be more important, was the venture to question the ideas of development: an idea that has so far been indisputable. Combining the term “development” with an adjective "sustainable" has resulted in an enormously important implication: it has pinpointed the impossibility of continuing with the currently indisputable idea of development, unique development without any adjectives. And not only are other forms of development possible, but there is a dire need for them.

Subsequent to this document, various other significant international initiatives were started, the more important ones being the United Nations Conference on Environment and Development held in Rio de Janeiro in 1992, and the World Summit on Sustainable Development, which took place 10 years later in Johannesburg.

Today, the concept of *sustainable development* has its stable place not only in the official language and documents of all international organisations, but also among all social *actors*, economic as well as institutional ones.

But great success on a communicative level has not brought any significant changes. Moreover, it seems that the very expression *sustainable development* has lost its original force. This has happened due to the overuse of the term in the ambience of events and situations that have little or nothing to do with its actual meaning. As a result, the risk of turning it into a semantically empty expression grows, with everybody agreeing, but not really understanding its deeper implications.

On the other hand, even if it is already semantically incorrect (*i. e.* due to overuse and abuse its primary sense has lost its significance), the expression "sustainable development" remains indispensable. And we think that what should be done is a kind of "semantic de-polluting", trying to use this expression only in the correct way and place, and at the same time criticising hard misleading instances.

All this requires in-depth theoretical examination of significance of the words used in the original definition, in this case, the exact meanings of "sustainable", "development" and "need"[1]. Otherwise, it requires moving from rather general ethical indications (*without compromising the ability of future generations*), to a concrete and precise action plan.

Hence, it is clear, especially considering the enormity of the topic, that the sheer number of fields covered and defined by this action plan will be huge; practically every aspect of contemporary society has to take advice. It is an enormous amount of work that can be brought to completion only as a complex and across-the-board process of social innovation, in which all participants have to play their part.

Our course of tackling these questions will start with a few notes on the concept of environmental sustainability before showing the extent of such a change that the sustainable point of view requires to be put into action. In the next couple of chapters, we will discuss the significance of this change, what concerns the aspects of well-being and the roles that the social participants in the production–consumption system are going to or could end up with; all in order to take some concrete steps towards sustainability.

---

[1] Theoretical discussion on the topic of *sustainable development* and its implications has created a vast bibliography. Just to mention some of the books that have been particularly relevant for this very chapter: Daly and Cobb (1989); Shiva (1989, 1983); Hawken (1993); Hawken *et al.* (1999); Sachs (1999, 2002); Marten (2001); Solow (2002); Brown (2003); Sen (2004).

1.2.1 Preconditions of Environmental Sustainability

The starting point of our discourse is the banal, but often forgotten observation that our society, and hence the lives of our and future generations, depends on the long-term functioning of the complicated ecosystems that we happen to simply call nature, on their quality and productive capacity (and in turn, on their capacity to produce food, raw materials and energy). And for us this has fundamental importance, that from this simple but important observation we can introduce the concept of environmental sustainability.

The term *environmental sustainability* refers to systemic conditions where neither on a planetary nor on a regional level do human activities disturb the natural cycles more than planetary resilience[2] allows, and at the same time do not impoverish the natural capital[3] that has to be shared with future generations. These two limitations, based on a prevalently physical character, will be aligned with a third limitation, based on ethics: the principle of equity states that in a sustainable framework, every person, including those from future generations, has the right to the same environmental space[4], that is, the right to access the same amount of natural resources.

Any serious consent to these sustainable requirements will expose immediately the actual division between the production–consumption system in contemporary industrial society and sustainability. The signs of this division are for example the unwise use of renewable resources (over-exploitation of some, like fish, and under-employment of others, like solar energy), the similarly unwise consumption of non-reproducible resources (fast exhaustion of many reserves and corresponding accumulation of waste), dispersal of a growing number of synthetic substances into nature, potentially noxious, unknown by nature and therefore non-naturalisable.

1.2.2 Ten Times More Eco-efficient Production System

It is perhaps useful to give an intuitive clue about the extent to which such a division is possible and to present a compendious image, which, despite inevitable simplification, allows us to gain an idea about the destination we have to head

[2] Resilience is the capacity of an ecosystem to overcome certain disturbances without irrevocably losing the conditions of its equilibrium. This concept, extended across the planet, introduces the idea that the ecosphere used by human activities has limits on its resilience that, when surpassed, give way to irreversible phenomena of deterioration.

[3] *Natural capital* is the sum of non-renewable resources and the environmental capacity needed to reproduce the renewable ones. But it also refers to natural diversity, to the amount of living species on this planet.

[4] *Environmental space* is the quantity of energy, territory and primary non-reproducible resources that can be exploited in a sustainable way. Also, the term *ecological footprint* has been used, which indicates the available amount of resources needed to live, produce and consume on a personal, national or continental level without surpassing the sustainable level (Wackernagel and Rees, 1996; Rees, 1992; Friends of the Earth, Wuppertal Institute, 1996; Chambers *et al.*, 2000).

towards, and about the route we have to take to arrive there. This image is based on quantitative information metering the reduction of resource consumption that will become necessary for all mature industrial societies in order to measure up to sustainability. In this way, we can estimate the extent of changes that in one way or another have to take place.

It is widely known that the impact of human activities on nature depends on three fundamentals: population, demand for a better quality of life and technological eco-efficiency, *i. e.* how the metabolism of the production system is able to transform natural resources into the requested well-being[5].

Starting from here, while bearing in mind the growing population, as calculated, and the growing demand of the emerging and developing countries, as justified, the third parameter surfaces – technological eco-efficiency – with a rather impressive result: the sustainable requirements are realistic, but only if eco-efficiency is increased 10-fold. In other words: we can consider as sustainable only those production–consumption systems that employ 90% less material input per unit of service than is actually accounted for in contemporary industrial society[6].

This estimation is approximate, but nevertheless valid to show the magnitude of changes that have to be ensured. Here, a vision of a society emerges in which one has to know how to live, and also how to live well, using only 10% of resources employed today in industrial societies. It is evident that the whole production–consumption system of this sustainable society will profoundly differ from the one we know today, so different that no partial modification, no incremental innovation of employed technologies, no *re-designing* of the existing system can really bring us there[7].

The sustainable outlook thus calls into question the whole model of development. In future decades we must be able to move from a society in which well-being and affluence are measured by the production and consumption of goods, to one in which people *live better* consuming (much) less and where economic development produces fewer material goods.

How exactly this passage can be undertaken is still unclear today, but certainly it entails a discontinuity that will touch every structural dimension: physical (material and energy flows) as well as economic and institutional (relations between social participants) together with ethical, aesthetic and cultural dimensions (criteria for

---

[5] This relation can be expressed with the IPAT formula: **I**mpact = **P**opulation × **A**ffluence × **T**echnology. Already in use in various ecological studies, it allows the estimation of different participants, especially the availability of goods and services (affluence) and the eco-efficiency of technologies (Ehelich, Erlich 1991).

[6] *Cf.* works of the Wuppertal Institut für Klima, Umwelt, Energie and the Advisory Council for Research on Nature and Environment (especially *The Ecocapacity as a challenge to technological development*, a study funded by the Dutch government); of the *Working-group on eco-efficiency,* which was promoted by the World Business Council for Sustainable Development (especially the final report *Eco-efficient leadership*, WBCSD, 1995, as well as in: Jansen, 1993; Schmidt-Bleek, 1993).

[7] *Cf.* Von Weiszacker *et al.* (1997); Schmidt-Bleek (1993); Braungart and Englefried (1992); Charter and Tishner (2001), The authors of this book have mentioned it in different works, especially in Manzini and Vezzoli (2002).

value and judgements of quality to legitimise it socially). It is equally certain that a long transition phase is ahead of us, or rather, transition has already started and now we have to manage it by trying to minimise risks and raise possibilities.

To avoid misunderstandings, we should bring in some specifications:

- The aforementioned discontinuity can arrive in a period of a few decades. Currently, among the people involved with ecology, this period is thought to be 50 years: a sufficiently long time to be realistic about such a complex transition, but sufficiently short not to seem too abstract to us. However, as it involves ours and our children's lives, it is within our immediate visual range.
- The possibility of foreseeing (and predefining) certain aspects of environmental sustainability does not imply any kind of historical determinism: the prerequisites of sustainability only affix some aspects of physical flows of materials and energy related to the metabolism of society. But the actual characteristics of such a society are still very much in the open: within the necessity for sustainability, a plethora of sustainable societies can be conceived in all their diversity. From this point of view, a mere assumption of the hypothesis about reaching sustainability does not predetermine the future at all.

In conclusion, the quality of sustainable society (or better still, of diverse sustainable societies that might emerge) depends on the patterns made by the transition. Although, as mentioned, we have to (and also can) try to foresee the physical preconditions of a sustainable society, we cannot foresee the social characteristics of a society that will occur under these conditions. These will emerge during the actual process according to the reactions of social participants, to the new emerging culture, to stabilising field lines and to the institutions created during this process.

The transition towards sustainability will be enormous and an articulated process of social, cultural and technological innovation, in which a quantum of corresponding options, various sensibilities and opportunities will find a place. Starting with observations and experiences about developments up till now, it is possible to draw a roadmap with diverse courses, which today seem reasonable.

1.3 Bio- and Technocycles

The common denominator to every path towards sustainability is the following assumption: in order to guarantee indefinite continuation of human activities and to avoid the worsening of ecological conditions, it is absolutely necessary for their ecological footprint to be lighter than the resilience of supporting systems; in particular it would be necessary to reduce to a minimum the extraction activities that impoverish ecosystems and all emitting activities that accumulate substances with unfamiliar characteristics and concentrations compared with initial terms[8].

As a resulting principle, these antithetical orientations are attainable: the biosphere (the sum of natural processes) benchmark and the technosphere benchmark

[8] For a clear vision of these problems *cf.* Capra (1996, 2002).

(the sum of all technological processes). The first orientation inclines towards maximal integration (achieving the biocompatibility of all technological processes: biocycles); meanwhile, the second orientation inclines towards the maximal non-interference of technological processes (achieving their isolation: technocycles).

1.3.1 Biocompatibility and Biocycles

The theoretical objective of biocompatible orientation is a production–consumption system based entirely on renewable resources, harvesting without exceeding the productive limits of the natural system, and on releasing into the ecosystem only biodegradable waste that is dispersed according to its naturalisation qualities (within the capacity to return component substances to natural conditions without excessive accumulation).

Biocompatibility orientation could be thus seen to be a naturalisation process of the production–consumption system. In fact, it leads to its recalibration until its existence appears to be a deviation, not a disturbance of the original natural cycles.

In practice, it entails a reorganisation of the production–consumption process as a chain of transformations (as biocycles), integrated as much as possible with natural cycles. From this perspective, both the quantity and quality of products and services offered by such a system are greatly limited. In fact, in the biocompatibility framework only those products and services that turn out to be efficiently compatible with existing renewable resources are allowed. And only on a par with the capacity of the ecosystems to absorb and demulsify the residues of the materials and energy employed.

1.3.2 Non-interference and Technocycles

An objective of the non-interfering orientation is an isolated production–consumption system that recycles and reuses all the materials employed, thus forming autonomous technocycles and not influencing natural cycles. It has to be mentioned immediately that the exact meaning of such an objective is unachievable even theoretically, due to thermodynamic laws that state that all hypothetical technocycles that are totally isolated from the biosphere are inconceivable (in any case, there will be exchanges of energy and increasing entropy). Nevertheless, this objective indicates an important direction and a viable path to be approached; in practice, it deals with creating industrial systems that emulate natural cycles, integrating all production and consumption processes within a (theoretically) closed production cycle, and in this way, approaching a minimum level of input–output between the technological and natural systems.

Non-interfering orientation could thus be seen in the light of the progressively growing artificiality of human activities, which inclines productive and consumptive processes towards greater autonomy from the natural resources and ecosystems they were based on.

Also in this case the amount and quality of products and services provided by such system are limited: the framework of non-interference allows only productive processes that are based on and compatible with the materials already existing inside the isolation and have optimised usage of energy.

1.3.3 Industrial Ecology and Dematerialisation

In principle, various sustainable production–consumption systems can be imagined, based on combinations between the orientations of biocompatibility and non-interference. Blocks of integrated bio- and technocycles can be derived that found the material basis for any hypothetical sustainable system, and what can be called *industrial ecology* [9]. An elaborate discussion on its meaning and implications, however, exceed the scope of this book; hence, only the difficulties of integrating two proposed orientations (biocompatibility and non-interference) will be observed. The first difficulty arises from their very definition: if biocompatibility means integration and non-interference isolation, then the conditions suitable for the first signify difficulties for the second, and *vice versa*. Biocompatible production–consumption cycles are by default consistent with activities dispersed along the territory (in turn consistent with diffused functions of ecosystems with which they have to be integrated). Meanwhile, technological cycles based on non-interference are reasonable only in situations of high density of productive–consumptive activities (the only efficient way to handle and integrate the materials and energy flows in isolation; besides, the only economically feasible way is to keep the distances short).

On the other hand, it is evident that the difficulties that arise from combining two orientations increase together with the scale of the system: the greater the employed flows of materials and energy, the greater the difficulties, either of making them biocompatible or of closing them in the industrial ecology.

To decrease these difficulties reducing the employed flows of materials and energy will clash with production processes that we have had so far, and with the final products (in the end, it is the amount and quality of products demanded by society that determines the magnitude of these flows). Operating in this way entails activating a process of *dematerialisation* of social demand for a better quality of life.

This expression signifies a drastic decrease in the amount (and in the *material intensity* [10]) of the products and services necessary for a socially acceptable quality of life, together with a collateral decrease in the flows of materials and energy employed.

---

[9] *Cf.* Hawken (1993, 1999); Pauli (1997, 1999); Fussler and James (1996); McDonough and Braungart (2002).

[10] *Material intensity* referring to products or services, implies the quantity of natural resources necessary for a service unit (*e.g.* transportation of one person for 1 km, 1 kg of cleaned laundry, 1 m^2 of painted surface *etc.*).

It can be effectively obtained by reducing the overall demand for products and services, heightening the overall intelligence of the system, and decreasing its functional material and energy flows[11].

1.4 Transition Scenarios

What has been elaborated so far explains some of the aspects behind the ecological sustainability and the roadmaps that theoretically could reach sustainability. But no practical guidelines for actual realisation have been provided. Hence, we have to move from the merely technical arguments outlined so far, to propositions related to other dimensions of this problem – its social, cultural and economical aspects.

As a first step in this complex field the following observation could be made: environmental sustainability is achievable without any traumatic phenomena only if such living conditions are foreseen and put into practice during the transition, and that the community perceives them as achievements compared with the actual living standards. In fact, this is the precondition to catalyse necessary energy for "voluntary" transition: nobody would ever choose to go in the direction of a destination considered worse than the starting point.

Thus, the possibility of putting into action a painless course towards sustainability is based on the capacity to envision it as a new and attractive quality of life that would also be a sustainable quality of life.

This intricate complex of problems gives the expression that no move towards sustainability is possible without keeping an eye on the social and cultural dimensions of this issue. This is only partly the case. For the same reasons, the so-called "soft transition" that we are using to outline a possible roadmap disallows any dramatic ruptures of economic systems. In other words, on the local and global levels, the overall socio-technological system has to change radically, searching for new economic assets that might guarantee it. Hence, no move towards sustainability is possible without keeping an eye on the economic dimension of this issue either.

On the other hand, at least today in industrial societies, the view on living standards and their growth has been connected to an increase in availability of goods and because it also means further employment of natural resources, the real issue lies in breaking the connection among the expected living standard, available goods and consumable resources.

Is it even possible to imagine such an event? In principle, yes. Breaking such a connection can be achieved if we employ a certain strategy that in turn consists of two complementary tactics: strategies of efficiency and sufficiency.

---

[11] In particular, heightening the intelligence of the system means to have the information and communication technologies turned into the nerve system of social metabolism and thus allow it to blend more easily with *industrial ecology*.

1.4.1 Strategy of Efficiency: A Radical Way of Doing Things Better

This strategy is founded on a basic assumption that thanks to advancing technologies it will be possible to maintain the current expectations of living standards (based on the aforementioned growth of consumable goods and services), and at the same time also reduce resource consumption. This objective is feasible when combining a severe *dematerialisation* of productive processes with strict application of *industrial ecology* principles[12].

Therefore, a strategy is predicted in which such reduction is possible owing to the technological discontinuity, *i. e.* due to a drastic rise in the eco-efficiency of the technological system; such an increase exactly allows social demand for life standards to be satisfied, without requiring substantial cultural or behavioural changes.

This is the way often advised by the less enlightened part of the industrial world: "continue to consume as always", is its suggestion. Ecological problems will be taken care of by experts, who will improve the efficiency of production systems.

If the objective remains to obtain a modest decrease in resource consumption, then this suggestion can be practical, unless we actually would like to take the sustainability issue seriously (and the actual necessary extent of reductions).

If we calculate the demand for products and services at a fixed rate, *i. e.* without influencing the socially accepted living standards, then there is no feasible way of transforming the productive system and of reducing its natural resource consumption to the required level. In other words, there is no wonder-technology, a hypertechnology that would at the same time allow the public demand for quality of life at today's level to be satisfied, and employ only a minimal part of natural resources.

1.4.2 Strategy of Sufficiency: A Radical Way of Doing Less

This strategy is founded on the belief that it is necessary and possible to actually have such cultural changes that would regard the reduced availability of goods as an increase in living standards. Therefore, what has to be done is to imagine a sharp cultural discontinuity within a substantial technological continuity. In fact, if the reduction of the consumption of environmental resources goes together with the decrease in available goods, then there is indeed no need to resort to any substantial technological innovation: true innovation in this case would be the radical changes in the idea of well-being.

---

[12] The topic of *eco-efficiency* was introduced to international debate back in 1993 by the WBCSD (World Business Council for Sustainable Development) (WBCSD 1993). For other opinions *cf.* Fussler and James (1996); Brezet *et al.* (2001).

There is no doubt that this strategy deserves great respect: anyone who could, within free ethical and cultural choices, reduce drastically the actual demand for material products, would indeed make an efficient contribution to environmental sustainability[13]. The only problem that remains is that when, considering the share extent and urgency of this problem, do they claim to expand this strategy to the entire human race at such short notice. In this scenario, the extent and depth of such a required social change prevents us from even thinking that it is possible to carry it out without coercion in the given time and with the ways necessary. And the risk here is of falling into ecological fundamentalism, *i. e.* leaving the camp of free choice for others, where what has to be done is ordered by who is considered to have the monopoly of truth.

1.4.3 Compound Strategy

Given the impracticality of these outlined "extreme strategies", it becomes evident that the only passable route would start from the shift that at the same time intensely invests in both the technological system and the social demand for living standards. A voluntary transition towards sustainability implies, in other words, a progression of suggestions that entail structural discontinuities and contemporarily handle all dimensions at every level of the society we have known so far. In such suggestions, ecological quality will emerge from compound strategies and compound interventions in different areas.

[13] *Cf.* Shiva (1989, 1993); Sachs (1999, 2002).

Chapter 2
Products, Contexts and Capacities

2.1 Introduction

Contrary to widespread and frequently observed social and political terms, the pursuit of sustainability is opposite to that of conservation. Conservation and regeneration of social and natural capital demand that we break away from the ways of living, consumption and production that dominate today, and experiment with new ways. If this experimentation falls short, if no experience will ever accumulate, if nothing is learned from the process, then will the true conservation continue? What will sustain the current, catastrophic ways of living, producing and consuming?

A social experiment of such a spectrum has to schedule everybody's participation. This is a phrase that abuses itself. Something that has always been said: everybody has to remember to switch off the lights after leaving the room and put the garbage into right can....Sure, but it is not enough. On the contrary, to understand participation to be a little personal effort to maintain one's house and planet is in the long run misleading. What everybody has to do is not, in fact, little incremental improvements on what normal way of living can be put forward, but instead, exactly as has been told, a change in the way of living. Radical remodelling that does not require any new obligations for success (the obligation to make an effort on behalf of Earth or planetary society). Instead it asks for a drastic rethinking of our conception of well-being. In order to reach it, we have to perceive as positive some customs and manners that today are considered indifferently or even negatively: fancy going on foot, eating home-grown fruits, being aware of the seasons, caring about things and places, speaking with the neighbours, committing oneself to the neighbourhood, admiring a sunset *etc.*

Are such changes possible? Will such an idea of well-being appear and spread, an idea that will not be considered a must, but as a new and good way of doing things? In principle, definitely yes, the notion of well-being is a human construct of the mind, thus it is replaceable. It just has to be understood how exactly it could happen, keeping in mind that this conversion will touch very deep aspects of expectations and functionality of contemporary society.

2.2 Product-based Well-being

The concept of well-being is complex and controversial. Let us start with repeating and articulating what has already been said: the idea of well-being is a social construct; ours is the consequence of relatively recent history (history of industrial society), which until last century was mainly located in Europe and subsequently in North America.

Therefore, the currently dominant ideas of well-being were created and consolidated in a particular economical and cultural context (of the industrial society back in the last century), when the understanding of the "limitations" was scarce and natural resources were taken for granted.

During the last century, two different ideas were created and consolidated: the traditional concept of *product-based well-being* (which during last 50 years has reached planetary proportions) and more recently, the emerging concept of *access-based well-being*.

2.2.1 The World as a Supermarket

At the beginning of the industrial era the combined development of science and technology brought to human beings possibilities never seen before: the possibility of *materialising* complex services in the form of products (a laundry service that materialises in washing machines, the service of playing music that becomes a radio or a record player) and the possibility of *democratising access to them*, producing them in increasing quantities at decreasing prices.

This unprecedented possibility brought with it an extraordinary spread of a particular form of well-being, a well-being that was recognised precisely in the possibility of individually possessing, showing off and consuming the products. And, moving towards more recent times and more affluent societies, the possibility of choosing between different options and devising, in this way, a personalised set of products.

In the framework of this vision, which we can define as the vision of *product-based well-being*, the emerging idea is that life choices tend to considered as choices among marketable goods and that, as a consequence, the freedom of choosing is coincident with the freedom of buying. Metaphorically, the contexts that best express this vision are the big *shopping malls*: places where there is the widest choice and, if we have the money to do so, the greatest opportunity to buy whatever we prefer.

The problem of this vision of well-being is that (as in the last two or three decades we have been forced to discover) it is intrinsically environmentally and socially unsustainable. And this for several interrelated reasons that have been widely discussed throughout these decades. Because today, 20% of the world population that lives more or less according to this model of well-being consumes 80% of the physical resources of our planet. Very banally, even if it deals with terribly dramatic topics, there is not enough ecological space to sustain all others: 80% of the world population that today consumes 20% of the resources, but legitimately tries to live better. To do it, they have to confront, or are forced to confront, the dominant ideas of well-being, the equation "well-being = more products".

If all habitants of this planet really try to have access in a similar way to well-being, the catastrophe would be imminent: an ecological catastrophe, if they had the access, or a social catastrophe, if a few had and the majority did not have this access (or rather a catastrophe of mixed characteristics, as unfortunately this may happen really soon, or worse, it is already happening).

The fact is that a dream of the world as a supermarket, connecting the growth of well-being with the accessibility to material goods, that inevitably cause the growth of resource consumption, is just not-extendible to all habitants of this planet. In the last century, when the economic and cultural context seemed to have forgotten any ideas of the existence of limits, the connection between the growth of well-being and resource consumption did not raise any questions. But today things are different. Now we have understood (or rather are forced to understand) that this connection brings along all kinds of problems, not only ecological ones, but also social, political, and finally economic ones. And consequently this entire vision of well-being becomes unthinkable.

2.2.2 The Paradox of "Light Products"

Since the discovery of problems that spread together with higher living standards ecological issues have permeated economic and political agendas. Many products have been redesigned, their eco-efficiency has been raised and they have in general turned out to be "lighter": their ecological "weight", "a footprint", left by their making, using, consuming and finally dumping, has been considerably diminished. In other words, refrigerators, washing machines, cars and all other industrial appliances, they all, one by one, consume fewer natural resources.

Seeing things in this way, comparing products and their individual "lightness" one by one, can easily lead to a conclusion that the modern production consumption system is going in the right direction, in a direction that will eventually establish sustainable conditions. Which is unfortunately untrue.

In fact, extending the observation on environmental qualities from single products to the system in its complexity will immediately explain that the situation has not been improved. On the contrary, the total amount of natural resources consumed keeps on growing.

Then one becomes aware that, when things become lighter, smaller, more efficient and cheaper, they are trying to change their constitution and multiply their numbers, develop for wider and faster consumption, converge with fashion (like wristwatches) or approach the ephemerality of throw-away items (like cameras). Similarly, the development of user-friendly interfaces that are supposed to ease otherwise tedious and difficult procedures tends to trivialise and eventually sprawl. For example the notorious "*push and print* syndrome", together with the proliferation of computers and text editors, has made editing and printing so easy that every document is now printed in countless copies, infinitely multiplying the consumption of paper.

Definitely a great and for many also tragic eye-opener of today has been the *rebound* or *boomerang effect, i. e.* a phenomenon, where certain technological decisions were believed to be ecologically positive, but turned out to generate even more problems when put into use. As a matter of fact, as the aforementioned cases demonstrate, every technological improvement that is intended to raise the eco-efficiency of products and services appears to become a new and "natural" opportunity for consumption for reasons that are inherent in the overall socio-technical system. This is accompanied by a substantial increase in the unsustainability of the given system.

The fact that nobody was able to forecast such a course of events springs from the technocratic culture that is dominant among observers who have always disregarded the complexity of the socio-cultural implications of technological innovations and, most of all, has never really recognised their systemic dimension.

Even if in hindsight it seems clear, it was a bitter discovery of inevitable, laid-out facts: the industrial mitigation of single product lines has brought and will continue to bring a cumulative and multiplying cause–effect relationship between mitigation and consumption growth, unless something else is changed.

2.2.3 Lightness as a Non-sufficient but Necessary Condition

The last statement of the previous paragraph should not be misunderstood: it only attests to what we have had to learn, *i. e.* that the product lightening on its own does not guarantee any greater sustainable effects among the final results. But it does not exclude the respective research into light products (*i. e.* low environmental impact throughout the life cycle) from being a beneficial course to be followed with all the energy available, even if we know that it alone is not a sufficient precondition for sustainable solutions, but more than ever it is an indispensable condition. Sustainable society can only be an outcome of structural changes, but in the end it has to be supported by a new generation of products, which without doubt have to be *light products* [1].

---

[1] *Cf.* Von Weizsacker *et al.* (1997); Berzet and Hemel (1997); McDonough and Braungart (2002).

2.3 Access-based Well-being

Let us return to events associated with notions of well-being and their evolution. Recently, at least in more developed industrial societies (but also in cosmopolitan strata of every other globalised society) a new idea about the quality of life has stood out and started to spread. This conception, which is related to the evolution of the contemporary economy towards a service- and knowledge-based economy, can be summarised in slogans "from material to immaterial possessions" and "from possession to access[2]". Against the background of this new economy, resultant ideas and manners, the old central position of material benefits in the conception of life standards becomes obsolete: well-being appears to be separated from the possession and consumption of a "breadbasket" of material benefits, and associable rather with available access to a series of services, experiences and intangible benefits.

2.3.1 The World Is Like a Theme Park

For a society that is currently saturated with material products, can immaterial goods appear to be something "that really makes a difference"? The most important aspect here is that currently, when different lifestyles are characterised by mobility and flexibility, the mere possession of goods seems to be too heavy and rigid a solution, an unbearably stiff habit for a background that wants to appear as light and flexible as possible.

In line with this vision, defined as *access-based well-being*, the quality of life is evaluated by the quantity and quality of available services and experiences. In this case, the *theme parks* are the most emblematic figures: a place, where one can choose, which experience to go through, and where everything is carefully designed to offer as exciting an adventure as possible, provided that one has the money to pay for the ticket[3]. This emerging vision, which is by the way already dominant in "more developed" societies, has proved itself, both on a social and on an environmental level, to be totally unsustainable.

Social restrictions are evident: the emphasis on a *user's experience,* elaborated in a short-sighted economic background, as is unfortunately common, results in a supply of *experience packages*, conceived and realised in the exact same way as traditional goods; as something meant to be consumed fast and in big quantities.

With the result of the metaphor of *theme park* being most appropriate to represent the new concept of well-being, *to live well* signifies a possibility to gain ac-

---

[2] *Cf.* Cova (1995); Pine and Gilmore (1999); Rifkin (2000).

[3] The vision of life as passing from one attraction to another, and the vision of the world as a theme park, is transforming after 9/11 into a world of gated villages, where everybody can feel safe from the dangers outside. But in the end, even gated villages can be seen as theme parks of security.

cess to a series of entirely pre-programmed and pre-designed experiences, as long as one has the money to pay for them. Meanwhile, others, who can not afford the entry ticket, are forced to stay out. It is absolutely evident that it is hard to come to another socially so unsustainable idea, as dividing the society between the *haves*, who are in the playpen, plunged into some programmed experience, and the *have-nots*, who stay behind the gates dreaming about entrance.

Now, on an environmental level the *access-based well-being* might seem favourable to the sustainable society at first sight. As a matter of fact, for a certain time period the emerging access-based society appeared as a promising opportunity. Because it breaks the equation *more products* = *more well-being* and introduces a new formula, *more information* + *more services* + *more experiences* = *more well-being*, it is a way of living that is based on relations and intangible goods (information and experience). Therefore, it was thought that if *access-based well-being* were to reach a dominant position, then the consumption system would become naturally and painlessly more sustainable. Unfortunately, this hope for an easy way to sustainability has found its disappointing end: the experience of the last decade has demonstrated the unsustainability of the freshly emerged notion about living standards, at least in this *spontaneous* manner as it appears today.

2.3.2 The Material Ballast of Information

We have to admit that, despite all efforts and fostered illusions, the motivations behind the growing consumption of natural resources are in abundance.

One of the fundamental reasons is nestled in the fact that the much vaunted information services, intangible goods and access to experiences on a large scale are inclined to increase, not substitute, the traditional consumption of resources (we can read all the books and listen to all the music we wish for, but we will not give up the air-conditioner or a new technological gadget).

The second reason lies in the fact that the substitution of material benefits for intangible ones will not cause an inevitable reduction in consumption *per se*: airfares, staying in hotels, and visiting theme parks do in fact demand a whole array of access-dedicated artefacts that evidently have a material essence and inherent consumption (to keep them operational).

The third motivation, perhaps less immediate than the first two, can be summarised thus; even though "information replaces materials", in the current economic and cultural context, "information brings along material ballast", and the increase in the information flow tends to create new consumption opportunities.

As an optimal explanation, the spread of efficient communication systems (from the phone to e-mail and video conferencing) allows people to be in contact from great distances, without moving themselves. All true, but nevertheless, at least until now, it has not diminished the overall demand for transportation. The answer to this *prima facie* paradox is in the end very simple: long-distance connections, permitting a list of long-distance contacts to be stabilised and maintained, also allow the number of such contacts to increase exponentially. But, on the other hand, every new business contact or emotional relationship that is obtained and stored in virtual form also requires for preservation "real" contact face to face between the interested parties. Because of this we can observe today the unseen urge to move across distances, no matter how long they are.

The fourth reason is related to the earthly side of "spirituality" and "experience", to the fact that the "spirit can't leave the body behind". Thus, also, the most spiritual interests generate resource consumption. If, for example, millions of pilgrims decide to visit the Jubilee and Pope in Rome, they presumably do it out of spiritual and immaterial needs. Yet a trip to Rome implies actual transportation and consumption that, according to their nature, require a huge amount of resources and energy. Similarly, it can be concluded that *spiritual needs* for novelty and exotics has turned tourism into a major industry – which is probably ecologically the most damaging industry altogether.

2.3.3 Service Orientation as a Pre-requisite of Sustainability

Taking into account the results of the service-based economy in their recently condensed forms, they do resemble analogical specifications on *product lightening*, and the effects relating to it. We have seen the demonstration that similar effects also appear here. And we have had to accept that service-based systems *per se* do not take sustainable forms. Also, in this case, it has to be included and underlined that even though they alone are not sufficiently sustainable, overall they remain an absolutely necessary precondition for sustainable development.

As a matter of fact, while it is impossible to foresee even one sustainable opportunity in a product-based economy or like-minded notion of well-being, the same ca not be alleged for service-based economies or like-minded notions of well-being. In the second case, in fact, several favourably developed and inherently *light* services can be imagined that can correspond to those notions of well-being and product systems that are actually economically and ecologically sustainable[4].

To outline how to bring products and services into line with the sustainable development of living standards, it appears necessary to return to and catch up with the topic of well-being and some of its more profound concepts. After all, not only do we need the new concepts *of* well-being (how against the contemporary and cultural background it could be possible to define new qualities of life), but also new concepts *about* well-being. That means discussion of the whole cultural background in which expectations about living standards take their form.

---

[4] *Cf.* Stahel (1997); Brezet *et al.* (2001); Mont (2002); Manzini and Vezzoli (2002); Manzini and Jegou (2003).

2.4 Crisis of Local Common Goods

All that has been done so far with regard to ecological topics and discussion about sustainability has still not changed the cultural background, *i.e* for what has interested us to discuss here was the idea that usual research about quality of life has had to be based on purchasing new products and services (thus still based on the equation: "higher quality of life = more products and services = higher consumption of resources").

To leave this *cul-de-sac* and learn how to "live better, consume less", we have to observe closely how the conditions of living standards are put together, and especially how the connection between the product-service supply and the quality of the all-embracing context we happen to live in is defined.

2.4.1 The Role of Common Goods

In the well-being model that is dominant today in industrial societies, there is a central role for individually purchased goods (either products or services) that has had an unwanted, but nevertheless sizeable, side-effect. Consequent to its very definition of well-being, this model underestimates all "goods" that are not reducible to a product on a market, that cannot be bought or sold: from basic substances like water or air, to social goods such as neighbourhood communities or citizen solidarity, up to more complex ones like landscape, urban public space or a sense of security. It is clear, that goods "without a market", usually referred to as *public goods* (meaning that their value benefits everybody and nobody in particular), make up a fundamental part of the *human habitat*, *i. e.* the quality of physical and social space, where human beings live and where the very products assume their meaning in the first place.

Thanks to this attitude of disinterest towards public goods, their *desertification* (abandonment and consequently degeneration) and subsequently their *marketisation* (transformation into marketable goods: bottled water instead of natural water, commercial centre instead of public square, private security firms instead of neighbourhood watch, *etc.*) have occurred[5].

---

[5] To this observation another one can be added, what will be desisted here, relating to the crisis of *contemplative times*. This expression refers to a timescape in which "nothing is being done", which has the fullest meaning. Most evident examples of *contemplative times* consist of watching a sunset or undertaking any mental exercise whatsoever. We can also assume that doing something (like walking, eating, talking with others) at a slower rhythm than considered normal by society will count as *contemplative times*. Traditionally, *contemplative times* played a great role in life and were sometimes considered as a privilege (in fact, in the past the poor had no time to contemplate). Today, things are a bit different and *contemplative times* are lacking in the lives of rich and poor alike. This continued disappearance can be reduced to two reasons. First, the *saturation* of life (the tendency to fill each moment with something to do, and more and more frequently, to fill with many different things to be done at the same time, as for example driving

2.4.2 The Sprawl of Remedial Goods

In the course of the 20th century we can empirically follow how the diffusion of marketable goods and services was parallel to the deterioration of common goods. On this observation we can develop a few working hypotheses.

The first working hypothesis asserts that there is a direct connection between the diffusion of marketable goods (however efficient they are) and the crisis of common goods and everything they freely contributed (in the economic as well as the ecological sense) to the quality of life.

The second working hypothesis claims that there is a direct connection between the crisis of common goods and the thriving of a new generation of remedial goods, *i. e.* goods and services that try to construe as acceptable living conditions that are much deteriorated.

The third working hypothesis, closely linked to the second, states that the thriving of *remedial goods* in turn causes further deterioration of common goods, and thus ends in a vicious self-propagating circle.

The obvious focus here is on *remedial goods*. Their common denominator is that using or consuming them does not improve the living standards (contrary to a washing machine, otherwise we would still doing laundry by hand) nor does it open up any new possibilities (contrary to the telephone, otherwise we would not know many people). What they do (or are trying to do) is simply restore to an acceptable level living standards that have already deteriorated.

The meaning of this phrase becomes immediately clear, if we take into account how exactly the deterioration of common goods creates a demand for remedial goods: as already stated, we buy bottled water, because local natural water is contaminated; we visit tourist paradises across the world, because the beautiful neighbouring landscapes have been destroyed; we fit the house out with electronic and remote controlled security systems, because there are no neighbours who would agree, discreetly and for free, to keep an eye on our place[6].

Once we have focalised the concept of remedial goods, it immediately appears that the whole category of these goods could be eliminated without any deterioration in the quality of life of their consumers, if it were possible to contextually eliminate the problem that created this demand in the first place, *i. e.* if it were possible to overcome the crisis of public goods that has given a reason for their existence. But this type of preventive therapy for environmental woes is obviously troublesome to arrange.

---

a car, speaking on the phone and drinking a soda at the same time). The second reason is the *acceleration* of time, a tendency to do everything at haste to open up a possibility (or rather an illusion) of doing something more.

[6] Not only this, widening a bit the definition of public goods allows us to claim that we buy and consume a growing amount of services and products just to refill the empty feeling left by the crisis of public goods that gave the capacity to entertain and be entertained, to communicate and to listen, that for the past generation was a common and widespread ability. For this reason it is hard to distinguish which goods are remedial and which are not, but we can surely say that a great number of modern products, from the television to the mobile phone, from videogames to certain articles of food, observed in this manner, constitute a great part of remedial goods.

2.5 Context-based Well-being

We have followed the crisis of common goods, we have seen how this complex phenomenon has created a demand for a growing amount of remedial goods, and we have understood how it all ends in a vicious circle: more consumption, more contextual degradation, more consumption (of remedial goods) and so forth.

If these observations are correct, it ensues that any sustainable concept of living standards has to break this vicious circle first, *i. e.* redefine the meaning of well-being, cut its causal relation to the availability of new products and services, learn how to live better, consuming less, as was said at the beginning, besides the problem of restoring the social and ecological ambient we once lived in, in particular, restoring the common goods.

Is it all feasible? Obviously nobody can answer this question today in a certain and exhaustive way. However, it is possible and necessary to create hypotheses, to try to find a way to implement them and observe the outcomes. In other words, it is possible and necessary to feed a social learning process, which, as we have already said, is our only hope of leaving the cul-de-sac in which we are cornered, and move towards sustainability.

Summarily, the working topic is: how to change society so that expectations in terms of well-being would be less dependent on access to new artefacts, and more dependent on the recognition of the quality of the physical and social ambient in which we live.

The question thus formulated reveals that the object of observation (and action) in the search for well-being is actually a sophisticated entity that consists of the two just mentioned components (new artefacts and environment), and that could be defined as *life context*. Thus, it would be correct to have a life context and its qualities as a reference, while seeking new forms of well-being. Therefore, sustainable well-being should be seen as *context based well-being*, *i. e.* well-being that takes into account the entire scene in which human life takes place[7].

Reference to life context can be a background against which to have the social discussion about well-being and where to place working hypotheses, as a first move towards changing the game rules, rules that would form a basis for new concepts *about* well-being.

Here, to take another step in observing the well-being, the concept of *context qualities* should be connected to the concept of *capacity development*. As a preface to the latter, we have to start from the dawn of modern times, especially from the connection between accepting the concept of well-being we have talked so much about, and of *comfort*, as it has appeared in industrial societies.

[7] Here, context has a double role of *operational environment* (physical and social space related to the activities that take place and acquire a meaning within it) and *window of possibility* (set of connections and opportunities that in a given place and time give the person the possibility to act in this context).

2.6 Well-being as a Development of Capacity

We have already seen that well-being models have until now been born from a diffusion of mass production and consumer goods. And in particular from the enthusiastic discovery, that such artefacts can be manufactured that are able to work for us, as modern mechanical slaves. From the memory of hardships of many everyday aspects during the pre-mechanised era was born an idea of *comfort as minimal personal effort*. As an answer to any question, this idea claimed that the best strategy would be *always* to minimise the required physical effort, attention and working time, and consequently also minimising capacities and competence that one has to bring into play.

The success of this idea can be easily taken for granted. And, in fact, this is what happened: in the end, who has not tried, at the first opportunity, to minimise the fatigue, time and physical stress required by difficult and pestering everyday problems? But human nature is not so simple and mono-logical. A legitimate desire to leave the hardships of many everyday aspects of the pre-industrial era and the annoying repetitions of others is not an all-encompassing desire, applicable in the same way to everything in life. Human beings may tend to be lazy and passive, enjoy being served, but they can also choose the opposite direction. They can find satisfaction in a job well-done, or rationalise which working strategy is most comfortable for them, and find out that the best scenario is to do some things by themselves (because finally it is the most inexpensive way, or perhaps because it provides the greatest liberty)[8].

Of course, this active and participating form of human nature is anything but an all-encompassing treat (*i. e.* it is not something that is always and only this way) nor is it the only ethically acceptable way of being (as the rhetoric about the value of working pretended in some regimes of most tragic memory). Human nature is contradictory; it provides a way of acting logically or of following different dreams – this is its richness. From here the possibility also arises of rethinking the idea of *comfort* as a form of well-being based on increasing one's capacities. But first it should be explained why exactly the *comforts* of today are unsustainable.

2.6.1 Unsustainable Comfort

Against this background of positive diversity and various ways of doing and being is going to be articulated the critique of the production system, providing services

[8] Articulating this proposal, we refer to the line of thought in the studies on living standards and well-being by an Anglo-Indian economist, the Nobel Prize winner for economy, Amartya Sen. According to Sen, well-being is not defined by commodities and nor are its characteristics, but instead "a possibility to do different things, using a given commodity or its characteristics. " (Nussbaum and Sen, 1993). This very possibility allows in most cases the person t
a well-being, providing him with an opportunity to "be" (who he wants to be) an
he wants to do) (De Leonardis, 1994; Balbo, 1993).

based on a monological idea of comfort, *i. e.* services that have, progressively, confined the formerly widespread practical knowledge, integrating it in mechanical tools and organisational systems, and that have, in doing so, drastically reduced the capacities of people and community to autonomously face and solve their everyday problems.

This thrust towards passivity and ignorance in the functioning of everyday life systems has cultural and philosophical implications, which shall be discussed. But they also have grave and immediate ecological implications: production–consumption systems based on user passivity and ignorance cause a growing expenditure of natural resources and energy. Furthermore, they push people towards giving up caring about things and the environment: from the maintenance of things, to the attention to the context's ecological qualities. In conclusion, direction towards passivity and abandoning responsibility is an ultimately unsustainable way of living and doing.

Now the possibilities of changing this direction have to be discussed, or rather if and how new opportunities can be offered to these people and these communities; if and how it could be possible to envisage a new technological system that is capable of also inciting their sense of sensibility, competence and initiative.

2.6.2 Disabling and Enabling Solutions

To effectively focus on most interesting aspect, let us try to outflank the current *disabling goods and services* with production-service systems that could actually provide *enabling solutions.*

If today the products and services are generally designed for customers considered as troublemakers, to passive customers, then the new strategy should focus on what the user knows and can and wants to do. And it has to supply him with *enabling solutions*: such product-service systems that offer the means to acquire what the user wanted to reach in the first place, using the best of his abilities. And, if this should be the case, also encourage him to get involved and join in the game.

To put it in another way: for the versatile human nature and all the economies to which they can adapt, it is necessary to wage the design and technological opportunities in a direction that empowers users and communities with the sense of initiative, collaboration and belief in one's ability to find a better way of living.

As far as we can tell, the future perspective of well-being based on common goods and the development of capacities is indispensable, but it can also be assessed that it is a very likeable prospect. Well-intended of course: it is not predestined, not a future that will arrive no matter how, but a possible pathway to effectively actualise what we are capable of.

2.7 The Forces Behind Changes

Today it is impossible to foresee how the manifested model of well-being can actually take place and evolve, but it is certain that its eventuality is based on the existence of social and cultural forces inside the society that could promote it. Do any such forces today? Yes and no, depending on how we look at it.

To explain it better: within the dominant tendencies, together with emerging ones, all that has been said and proposed (a well-being that is based on context, regeneration of common goods and local resources, and concurrently, considering a well-being as the development of capacities) appears as a bundle of good intentions without any consistence, and is thus doomed from the beginning. Ideas about context clash for example with the crisis of ideas about places, and with consequent indifference about the very space that is our living context. And the proposal about regenerating local resources and common goods clashes with the rooting, the individualisation and colonisation of all available living space by the market. Finally, the consideration of well-being as the development of capacities appears contradictory to the inactive and passive tendencies, banalisation, or what might be called the *McDonaldisation* of life (from cooking to organising a party for the kids, transformed into standardised services).

However, if now we are not interested in the statistical weight of the considered phenomena, but in their potential to change the future, observing the complexity of social dynamics, the emerging picture may change.

As a result of a more precise observation, certain individuals and communities can be recognised that head in a different direction from the aforementioned one: individuals and communities, for whom the concepts of *context* and *common goods* are in fact known and field-tested territories[9].

However, this observation *per se* would not say anything very meaningful about the role of these groups or about the socio-cultural dynamics they represent. In the end, groups of people who have thought and acted in a way analogous, at least on first glance, to our description, have always existed, always in minorities, and therefore irrelevant. So, what is new today?

There is something new and very significant indeed, that is, their way of proposing their ideas today. As a matter of fact, in modern society the study of ways of sustainable living and developing sustainable systems exists for a wide range of reasons: ideological, just as in the past, but also for functional and economic reasons. Many who change their attitude towards well-being and act accordingly do it because it seems more logical and comfortable with regard to their specific living conditions (trite, but significant choices like not having a car in a congested city, but using public transport; eating organic food, as long as it can be easily found; promoting new ideas about the neighbourhood and offering the community their services to deal with difficulties and high living costs among young couples or lonely pensioners).

---

[9] *Cf.* Moscovic (1981); Rheingold (1994); Balbo (1993); EMUDE (2006); Meroni (2006).

Right now, a great many people find themselves in these situations; in the future, due to ecological and urban pressure, this number may be far greater. Besides, increasing the connectivity and maturation of a network-based society, together with the growing costs of natural resources, can turn some of these choices, which today are in a minority, into "normal" decisions, *i. e.* accepted because they belong to the primary economic and functional criteria.

The radical difference from the past is that the socio-cultural and technological framework in which innovative ideas of well-being are proposed and developed is totally dissimilar to that of a few decades ago. Even if, as we have seen, on average, this framework is still very negative (as the dominant trends are not moving towards sustainability), it nevertheless also seems able to offer some very promising opportunities. Thus, the combination of the emerging networked organisations, together with "the discovery of limits" (and all that follows from it), can make way for "new ways of living" that are coherent with sustainable principles[10]. A phenomenon that, as we tried to show in this chapter, could take place not only because people will consider its implications to be ethically correct, but also and especially because they will see this as the opening up of new opportunities.

[10] For general context *cf.* Giddens (1991); Augé (1992); Bauman (1999). For new organisational forms spread via the internet *cf.* Kelley (1994); Benkler (2006).

Chapter 3
A Social Learning Process

3.1 Introduction

Transition towards sustainability will be a social learning process that will teach us, progressively, through mistakes and contradictions – like every other learning process – how to live better, consuming (a lot) less and how to recreate the physical and social ambient we live in.

This phrase summarises effectively all that has been said in the previous chapters. In particular, this sentence claims that in order to avoid a social catastrophe, this transition has to be effectuated by positive choices, and the results of the transition have to be perceived as improvements in living conditions.

The second chapter showed that new concepts of well-being have to be created and spread in order to secure such a possibility. The following chapter will focus instead on a parallel transformation that has to invest in the *production–consumption system*, where such ideas of well-being can take place, and in the role that the different social actors involved are supposed to play.

3.2 The Production–Consumption System

The term *production–consumption system* refers to a complex social and technological system in which socio-culturally and economically available natural resources are transformed into a supply of products, services and public goods that responds – or at least is supposed to respond – to a demand of well-being in the given society. It is all a great interactive network among people (consumers/users), organisations (public institutions, private and social enterprises, civil society institutions) and territorial resources (natural and social capital, the input of *production–consumption processes*).

In the face of this complexity it is natural to ask how the radical systemic changes that the transition towards sustainability requires will take place, and who will design them. Some initial answers to these questions have already been given in previous chapters: the design role in this transition is not to be seen as the updated delirium on the design power that in the past has characterised occidental culture (the belief that the world could be totally redesigned by a subject – provided that he could be strong enough to control all the variables). It will appear instead as a set of partial contributions to a huge, complex and probably contradictory phenomenon of social innovation that will involve all social participants[1].

It must be added that because of its complex and contradictory character, at any given time, the production–consumption system entails different development hypotheses: those dominant today and those capable of giving new answers for new social demand. In other words: it is the same socio-technological system of production–consumption that in its complexity entails a field for cultivating new concepts of well-being and an incubator for the forces behind changes that could auspiciously sustain and defend them.

While referring to different social participants, it is obvious (but perhaps useful) to remember that every one of them operates according to his or her own point of view, follows his or her own interests and his or her personal value and quality judgements. From the other side, since between these different social participants are established complex interactions (*i. e.* reciprocal connections, which influence everyone's behaviour, which in turn is influenced by that of others), each behaviour is analysed against the behaviour of all and against the social rules within which everyone is operating.

In the following paragraphs we will develop some observations related to the four principal participants of society mentioned: consumers/users, producers, the public institutions and designers.

3.3 Consumers/Users and Co-producers

Let us consider people as consumers of goods and users of end-products and services, *i. e.* as consuming what the productive system produces. Therefore, they are the social participants, whose demand is supposed to determine what is produced and supplied.

In reality, as we know and have seen in previous chapters, the relations between demand and supply are a lot more complex and the reasons behind a certain structure of demand and supply depend on another intrinsically complex set of factors (because of this it is more efficient to describe these phenomena together as a production–consumption system and not as a production system). In particular, following observations could be made:

- The role of a person is never just as a consumer or a user. Every act of consumption requires active participation by the consumer/user: purchasing and preparing, which could be a simple task (buying a pre-prepared and deep-frozen meal

[1] This phrase summarises the debate of last couple of decades on designing in complexes *cf.* Bocchi and Ceruti, 1985; Ceruti and Laszlo, 1988.

and throwing it into a microwave; calling a taxi to let it to take him wherever he wants) or a complex one (buying ingredients needed for home-made pasta; going to work by bicycle). Anyhow, nobody is just a passive consumer or user. Everybody is asked for participation, which depending on the case, might demand more or less time, energy, attention or necessary skills, but it is still there.

- What is produced or supplied is not determined solely by the demand of a given moment, but also by either internal factors of a production system (*e. g.* unseen opportunities created by technological innovation or globalisation, or, on the contrary, inertia of established norms) or external ones (*e. g.* standards and incentives dictated by society according to emerging social and environmental problems). In the end, a person who was supposed to be the end-user finds himself with products or services for which he has not expressed any demand, and at times has trouble understanding their utility.
- At any given moment the demand for products and services within the given socio-technological system has very diversified characteristics. One of those is connected not only to the personal taste and individual character, but to the vision of quality and well-being. From this point of view, even if there are a number of demands according to variations of the same way of living and doing, others still exist that refer to other radically different ways of thinking and living. The ways of living, that today are in a minority, but to us (as stated in the previous chapter and as we will resume later on) may be of great significance and interest.

For these reasons, which offer us a more articulated and complex vision of what is meant when speaking about consumer/users, we should wonder what would be their current (and possible) role in the transition towards sustainability.

3.3.1 The (Potential) Strength of Consumers

Let us start from this statement: each individual, deciding how and what to purchase and use, gives a legitimate reason for the existence of that product or service (and therefore also for the origin of the environmental effects linked to its production, use and disposal).

In reality, we have seen that this is not entirely true and that his choices are shaped by a variety of factors beyond his will: the idea of well-being that applies here is not a free choice, but depends on the socialisation he has had. And, above all, the structural conditions in which he lives, and the alternatives offered within this framework are dependent on past choices made by others (the choices of other consumers/users before him, of the public institutions and producing companies). Nevertheless, it is certainly true that ultimately the production system is justified, because it responds to a social demand: the demand for the products and services that this production system is able to offer.

This insight results in the issue of critical consumption (*green consumerism*), in the discussion on how feasible, widespread and efficient it might become.

3.3.2 Critical Consumption

Critical consumption is an already established phenomenon and has played an important role in placing the previously whistle-blowing environmental issue among the topics of economy and markets. In the most recent environmental policies it tends to be promoted, thus increasing the possibility that users will make such choices of consumption, which will affect the market (and consequently the environmental quality of the products and services).

These initiatives are designed to raise the awareness of the consumers/users and their abilities to make choices. At the same time it should give them a real chance to make these choices: which means that there must be the alternatives that could be easily assessed by the consumer/user.

The *Eco-label* (a mark of environmental quality) and the *Energy-label* (to label refrigerators and fridge-freezers according to their energy performance) are, for this topic, the important instruments of environmental policy, and by providing the user with correct information, act directly on the mechanisms of the market. The labelling of *fair-trade* products is another significant example (these issues will be discussed later, together with the public institutions and their environmental policies).

The practice of critical consumption is not confined to the topic of the role of consumers/users in the transition towards sustainability. As a matter of fact, we have seen how and to what extent the system of production and consumption has to change.

And how and to what extent must the idea of well-being related to and partly created by the production–consumption system change? A change that cannot be obtained simply by the fact that ecological and fair-trade products can be found on supermarket shelves.

The way of living has to be changed, and not only the choices of consumption. In a way, such behaviour becomes consistent with concepts of well-being that recognise the value of the context, the social and ecological public goods. Where people are seen with capabilities that should be enhanced and developed, in order to enable them to play an active part in resolving their problems and recreating the quality of the environment in which they live (Chap. 1). What can we say about this change and about the actions that could promote it?

3.4 People as Co-producers

To respond to this question, we should start from the opening of this paragraph: each obtainable result requires the participation of the user/consumer. If this is true, it follows that each product and service, in the implied or explicit form, refers to a model of consumer/user characterised by its willingness to be a part of obtaining the result.

As mentioned in the previous chapter, during the past few years a model of user has evolved and become established who is principally concerned with reducing the effort, time and attention required from him. We can now claim that the marketing, with its insistence, has created a place for an increasingly lazy and disinterested user. The consequence of this vision has been a trend towards what could be called the "throwaway planet": a world in which the user is offered the "luxury" of obtaining functional results and sensory excitement just by paying the smallest amount of attention and making the minimum effort.

Besides, all other possible ways of assessing the "value" of degrading the quality of products and the experience brought by a widespread throwaway planet, the environmental consequences of this behaviour are evident and demonstrate clearly that this vision of consumer/user, which is the basis of this throwaway world, should be amended.

In the sustainable perspective, in fact, people should be considered as an active part in the processes of caring about things, the public goods and the environment in general. For this they are seen as people with the ability and willingness to struggle and act as co-designers and co-producers of the production–consumption system that concern them and the places where they live[2].

3.5 Active Minorities and Auspicious Cases

All that has been said creates another question: if and how the move from the model of consumer/user to co-designer/co-producer is going to take place (or could take place). As already said in the previous chapter, the prevailing behaviour is all but promising; on the contrary, unfortunately, the dominant dynamics give way to an increasingly unsustainable lifestyle. Things change if we focus on lifestyles that are adopted by minorities, but which have given birth to promising social innovations, *i. e.* lifestyles that have taken concrete steps towards sustainability.

As a matter of fact, we can see that in all industrial societies the number of people is growing who have found the desire, energy and ability to focus on some results and invent original ways of achieving them. Groups of people with initiative, called here the *creative communities*, which, for example, are experiencing new lifestyles by pooling residential services (and creating different forms of co-housing). Associations that promote the neighbourhood life by creating conditions for children to walk to school (as in the case of the *Foot Bus*), or stand for alternative forms of mobility (walking or cycling). Communities that implement new social services for the elders and parents (cohabitation of elders and youngsters, day care centres promoted and managed by mothers). Groups of people who build new food supply networks and prefer the local producers of quality or organic

[2] The idea about people with capacity comes from the aforementioned thesis by Amartya Sen (*cf.* Nussbaum and Sen, 1993; De Leonardis, 1994; Balbo, 1993).

food (like the *Slow Food* or *fair-trade* purchasing groups). The list of promising cases could continue[3].

As far as we are concerned, in the light of these established capacities, the issue of new sustainable solutions has to be considered important: new available social resources thanks to design that has been equally effected by the specialists (technical and political) and ordinary citizens (consumers and users); like a meeting of experts, where one side bring their traditional competence, while the second comes with their direct knowledge of the problem, with creativity and initiative to generate new ways of fighting against the problem. A two-sided active process of co-design and co-production, open and dynamic, capable of making the most of all the social resources available (Box 3.1).

Box 3.1 Creative Communities and Collaborative Network

The *creative communities* are groups of people organised to obtain a certain result, to solve a problem and/or to open a new opportunity. They involve *the first hand activation* of the interested parties. And thus, as a whole, they imply *the capacity and desire to do* (alone and/or exchanging mutual aid), rather than to ask (as in the tradition of the users of social services).

These initiatives are self-created by the citizens themselves. Or often born from the fruitful encounter between social innovation and social or private enterprises. Anyhow, even when institutions or companies are involved, these initiatives are created and act from the bottom, on a local scale, and, all together, are the creation of distributed creativity and entrepreneurship.

This characteristic connects them to other phenomena of contemporary society, in particular, to new activities that are growing in the new internet economy and collaborative networks.

The starting point of this new phenomenon is well-known: the idea of Open Source in the ambient of information technologies has shown its capacities with peer-to-peer organisations that are able to attract a large number of participants, give them a common goal and develop very complex and difficult projects: the

[3] The examples (and relevant comments that have been presented) emerge from research carried out by the Faculty of Design and of the Department INDIGO, Politecnico di Milano, in connection with other European universities, with research centres and the UNEP (United Nations Environmental Programme). Of particular importance was the collaboration with many schools of design in the world: a partnership that was, operationally, embodied in a series of workshops on the subject of daily sustainable life in countries like China, Korea, Japan, Canada, USA, Brazil, India, France, Finland and Italy. From this collaboration was born the *Catalogue of promising cases,* an exhibition *Sustainable Everyday* (La Triennale, Milan, 2003) and a book with the same title (*cf.* Manzini E and Jegou F, 2003). Subsequently, the research Emerging User Demands (EMUDE) has been developed: European research funded by the 6° Programma Quadro, conducted by various research centres and design universities, and coordinated by the INDACO Department of Politecnico di Milano. Many of the cases that are mentioned in this book can be found at the address: http://www.sustainable-everyday.net/cases.

movement that was built around the topic of free software has been able to generate products and services sophisticated enough to compete with the great monopolists of the sector.
We are interested in how this event started and then evolved. The same reasons that have brought thousands of people to work on a collaborative project of free software, the same organisational forms that allowed them to realise it, have been applied in other fields, and created, for example, new forms of distributed knowledge (the best-known example is Wikipedia, an encyclopaedia written voluntarily by thousands of authors and in a few years has become the largest encyclopaedia in the world). Or, what is even more interesting, unseen forms of organisation and socialisation have been created (as in the cases of *MeetUp, SmartMobs* or the *BBC Action Network*: platforms with common feature to gather together individuals concerned about the same topic, to collect the critical mass required to do something, and act accordingly. Let it be sharing a bus for a trip, organising a party or putting together teams of volunteers to clean a forest or a beach).

3.6 Enterprises and New Forms of Partnership

Out of all the social participants within the production–consumption system, companies hold most of the resources in terms of knowledge, organisation and capacity for initiative. They therefore play a central role in the promotion of sustainable processes. But first, such a possibility has to be compared with the economic efficiency and competitiveness. Which is no easy task.

Until now the encounter between companies and environmental issues has been on the following terms: how can a company remain competitive despite the new ecological constraints? This is, of course, a very legitimate question, but the way in which it portrays the issue is intrinsically weak: the orientation towards sustainability is proposed as a problem to be treated alongside other problems that very often are perceived by the businesses as being much more urgent and pressing.

In recent times, however, another, more interesting way has evolved: how can sustainable processes make the company more competitive? This means that competitiveness can become the mobilising factor. The best design and entrepreneurial resources in the search for sustainable solutions?

3.6.1 Producing Value by Reducing Consumption

As said at the beginning of this chapter, the transition towards sustainability is a learning process that will teach us how to live better, consume (a lot) less and how to recreate the physical and social ambient we live in. This phrase may also be remade specifically for companies. Then it becomes:

For companies, the transition towards sustainability will be a learning process that will teach them "to live well" (i. e. to produce value in a competitive way) offering their potential customers and society as a whole solutions to guarantee, or improve, their well-being by consuming fewer resources, and by regenerating the quality of their environmental and social contexts.

It is quite clear that for most current companies the cultural and operational reorientation necessary to successfully devise, develop and supply "solutions for better living by consuming less" will be neither easy nor painless. However, as with every other process, even the transition towards sustainability provides great opportunities and not only problems.

In fact, if is true that companies and their industrial activities are the biggest cause of the environmental problems confronted today, it is also true that, within this production–consumption system, they have the most advanced knowledge base, organisational resources and initiative. As said before, the companies know, more than any other social actor, the field of possibility (*i. e.* they know what is technologically or economically achievable) and have the operational capacities (*i. e.* the organisational and business resources necessary to transform the socio-technological structures within which they work). In short, companies can and have to become the most active *agents of sustainability*[4].

On the other hand, to operate successfully in this direction, companies have to change their *business-as-usual* attitude. On the contrary, the amplitude of required changes shows that one has to take two steps away from the *business-as-usual* attitude. First, how to actively use the changes (evolution towards service- and knowledge-based economy) and in this context also how to conceive *new business ideas*. Meanwhile, the second, more difficult step will teach how to follow the new directions of sustainable development and accordingly create *new business ideas*[5].

3.6.2 New Methods of Running Business

Now, for this two-step stage, one has to think anew what exactly business activities are and what is their outcome. In fact, their only chance to "create additional values while reducing the quantity of material products and improving the contextual values" lies in recognising the new concepts of market, society and their position within.

In particular, the whole meaning of *business* has to be reconsidered, within a wide structure where traditional businesses encounter other types of enterprises,

[4] The possibilities of companies acting as sustainable agents have been studied by many researchers. *Cf.* Hawken (1993); Pauli (1997, 1999); Braungart and McDonough (1998); Hawken *et al.* (1999).

[5] *New business ideas* that are based on the realisation of a new production system. A system of products, services and communication that allows the company to successfully develop its activities in new ways.

e. g. social enterprises, and where a wide variety of social participants are creating new production systems and unforeseen methods of creating revenue[6].

The ability to envisage new productive methods to create revenue involving different parties and combining social innovations with technological and economical innovations; all this could be the field in which companies could, auspiciously, become agents for sustainability.

The practical and operational definition of this field is outlined by two complementary strategies discussed in the first chapter and by their application in stages most agreeable to companies: *eco-efficient system research* and the *development of new solutions* provide an instrument to confront, with a sustainable approach, some important problems emerging in contemporary society.

3.6.3 Eco-efficient Businesses

As shown in the first chapter, the transition towards environmental sustainability can be described in technical terms as a process of de-materialisation of productive–consumptive processes: reducing the maximum amount of energetic and material resources needed for any given result. Or rather, for higher environmental sustainability it is necessary to increase the eco-efficiency of productive systems, disconnecting the value curve from natural resource consumption.

This general indication has been historically transposed to companies in a rather immediate way: it was necessary to re-organise the operational stages of their respective products in an ecologically efficient way. Whether it was a fridge, a car, a computer, a bottle of milk or water, under discussion was the maximal reduction of natural resource (energy or material) consumption that occurred during the production or utilisation stages.

In the recent past, much has been done in the field of *product eco-efficiency* and, as was written in the second chapter, the appliances of today are a lot more eco-efficient than those of 20–30 years ago. What concerns single products is that it can be claimed that a rather effective disconnection between the value generated and the natural resources consumed has taken place. But we have also observed that this strategy has fast shown its limits. Even if the *ecological footprint* of single products has been greatly reduced, resource consumption by industrial countries as a whole has continued to increase. This has happened because the growth of the overall quantity on the market has been greater than the growth of individual eco-efficiency (per unit produced). In short, the goods are a lot more efficient, but there are a lot more products than before.

From the first series of experience it has been concluded that product eco-efficiency is necessary, but not sufficient, to reach sustainability.

---

[6] *Cf.* social innovation initiatives where companies act as "process moderators", where they support with the technological, information and organisational resources, but not with finalised products; *cf.* Manzini *et al.* (2003).

As an answer to this realisation a new direction towards sustainability has been proposed by many, where the focus is no longer on single products, but has shifted to the eco-efficiency of product–service systems. And this in the well-motivated belief that environmentally optimising whole sets of products, services and productive processes instead of single goods would give greater results in eco-efficiency:systemic eco-efficiency, to be precise.

3.6.4 From Product to System Eco-efficiency

Therefore, the problem of eco-efficient businesses is increasing attention to wider product–service systems. There are several guidelines that can help to reach (and have helped to reach) this goal[7]:

- Deem the respective system to be *a product life cycle*: a product is designed within the framework of its different – production, use and consumption, disposal – stages.
- Deem the respective system to be a set of *symbiotic processes:* different productive processes are designed to use the waste of one process as an input in others.
- Deem the respective system together with other products and services to form a *solution to the problem:* designed to obtain a certain goal, these products and services are considered as one system.

Theoretically, all three approaches should be integrated in order to have any significant success in the right direction, *i. e.* foresee (new) product–service systems, designed within the life cycle framework, assessing every related productive stage, evaluating possibilities of integrating them symbiotically into other processes, and all from the perspective of *industrial ecology.*

3.6.5 Looking for New Solutions

Previous paragraphs proposed that the expression “eco-efficient business as promoter of systemic eco-efficiency” could offer for companies an approach to operating as an agent for sustainability. Now we should introduce a second approach, the one which we have referred to as proposing new solutions, solutions, that could be at the same time concrete steps towards sustainability and solutions to some urgent problems of contemporary society, *e. g.* demographic changes and increased demand for services by the elderly, young families and multicultural communities; local issues and demand for new development models; increasing differences between the “South” and the “rich North” and the call for solutions that could overcome them. The common denominator in all these problems is that,

[7] *Cf.* Pauli (1997, 1999); McDonough and Braungart (2002); Stahel (1997); Mont (2002).

given their size and complexity, they cannot be satisfyingly solved by the business-as-usual or other current approaches. Dealing with them requires totally new methods. And here the sustainability topic (in all three dimensions: environmental, social and economic) can provide some useful directions.

On the other hand, having a positive approach to sustainability brings along immediately an introduction to social sustainability side by side with environmental sustainability. Not only because social sustainability is just as much or perhaps more urgent than environmental sustainability, but because some issues do not have any feasible solutions (therefore they are not economically sustainable), if they were not based on recreating social context and do not conclude with answers based on solidarity and participation (problems related to taking care of the elderly constitute an emblematic case here, where even traditional social services do not offer any more feasible solutions).

3.6.6 Starting from the Results

In the end the final instance – consumers and users – do not ask for the products and services, but for the results they enable them to obtain[8]. But results can be obtained in different ways, combining various products and services (and various acts of personal participation by the user himself). It is self-explanatory that every combination has different social, economic and environmental characteristics, and also different relationships with deeper dynamics that drive contemporary society.

It is about being able to operate in this contradictory context focusing on and developing different combinations of products and services that are sustainable and at the same time recognisable and appreciated by enough people to also be commercially successful (or at least make the application economically rational).

In order to develop such a design capacity one has to define a strategy capable of investing in the whole business. A strategy that could mean working in a new way, creating partnerships that were unthinkable before, and considering not only the technological innovation, as companies have always done, but the social innovation too.

3.6.7 Business and Social Innovation

Finally, it is noticeable that many auspicious solutions for the problems covered here have not been discovered in the laboratories or research centres, but have emerged among individuals and communities, from their capacity to "invent" new

[8] This statement has its theoretical basis in the writings of the aforementioned Amartya Sen (*cf.* footnote 2 in this chapter and the footnote 8 in the previous chapter) and is operatively extrapolated by Walter Stahel and Oksana Mont (Stahel, 1997, Mont, 2002).

ways of acting and being. A diffused social innovation that, if recognised by companies, could indicate a new demand for products, services and new directions for technological development. *Promising cases* of social innovation, as we saw in the paragraph about new consumer's roles, could be the incentives for system innovation and the benchmark for the development of a new generation of intrinsically more eco-efficient products, services and production–consumption systems.

Let us consider the same simple examples in the aforementioned paragraph: the experience with co-habitation services could be the threshold for new instruments of unprecedented residential functions. Solutions that make available healthier and natural food or create direct relations with the producers can offer a new shape for the whole network of food production, delivery, and consumption. Examples of localised or auto-production can conceive new production processes and new products meant specially for these kinds of activities. Experience with alternative means of transport can make place for new kinds of public transport.

If interpreted correctly, these cases of social innovation represent new specific demands for product–service systems to which the companies should answer[9].

Box 3.2 Systemic Approach in the Era of the Internet

Systemic thinking and operating mean by definition moving up one level in complexity, and has always been so. But confronting this complexity today is a bit different from the last few decades for two reasons. First, the usual technological (and technocratic) approach has been replaced by a socio-technological approach; second, because the working environments have moved from *low* to *high connectivity*, as we have entered into the era of networks.

The first reasons could be summarised as follows. The traditional way of dealing with systems was to define them in technical terms and then try to manage them. This approach never went too far. Today, this approach has been overturned: the system is defined on the basis of what is thought to be operatively manageable.

At the same time, we know that, to have a manageable, and therefore, designable socio-technological system, convergence has to take place among all the actors involved, *i. e.* all of them have to share the same idea about the results to be obtained and the way to do it. The practicability of the systemic approach is traced back to the creation of *result-oriented partnerships*.

The issue of creating result-oriented partnerships is a crucial breakpoint for all enterprises (or more generally for every organisation) that actually want to adopt a structural strategy towards sustainability. This issue entails taking into

[9] Not only do they propose another relative demand for the general reorientation of the production–consumption system. The organisational models and some business models offered by some creative communities could indicate for research and development new *organisational and technological platforms:* service infrastructures for residential structures and polycentric production; an electricity grid based on distributed production; structures for incubating new complex partnerships; supporting systems of newly evolved participatory democracy.

account the interests of all social participants (participants in designing sustainable solutions that are not just the companies, but also institutions, local entities, voluntary associations and the final user).

Another characteristic aspect of the contemporary systemic approach is the requirement to enlarge the research field and engage the solution in the framework of current technological, economical and social transformations. Because all that today has been learned on the subject of the networks about supportive infrastructures and organisational metaphors has rendered possible new operational methods.

In particular, in the case of the systemic approach for design and production, it can be observed that networks allow us to move on from the notion of systems as great isolated buildings to be built from foundation to roof (as was seen during times of low connectivity) to a more sophisticated vision of systems as being part of larger and more flexible networks. In this new conceptual and operative framework, the topics of systemic eco-efficiency and of the management of complex solutions have to be attentively re-considered.

3.7 The Public Sector (and the Rules of the Game)

The public operator, meant here to be a sum of all entities in a position to speak for society, has, at least in theory, the possibility to set the game rules. Therefore, it has a fundamental role to play in determining the timing and method of transition towards sustainable society in general and the sustainability of the production–consumption system in particular.

It is obvious that orientating the public and private choice towards sustainability is furthered or restrained by general political, regulatory and economic conditions. It is also self-evident that in principle the role of the public sector is to create more favourable conditions, to be more precise, to activate a virtuous circle amidst individual and private sector behaviour, so that competition among companies would shift more into the field that favours solutions leading towards sustainability. In short, the public sector has to facilitate and promote this great social process of learning we have spoken so much about.

For a better understanding, let us take a quick step back before continuing with more general topics.

3.7.1 *Facilitate the Social Process of Learning*

If transition towards sustainability could be seen as an adaptation-through-learning process, then what is necessary above all is to develop the ability of a social and productive system to understand the feedback given by the environment and amend itself accordingly, and from here on, the capacity to understand the feedback and to employ other transformations, *etc.*

Thus, discussing the transition towards sustainability means also discussing the timing and methods of this process. It is evident that the later we understand and react, the more traumatic the eventual changes will be. It is also clear that bigger the scale of phenomena we are supposed to perceive, the more far-flung, the more difficult it will be to handle (and probably more dangerous for individual liberties), the correspondingly bigger the authorities to engage the response have to be.

So there are good reasons to confirm that the most auspicious solution on a social and political level is the one that makes use of the feedback as easily and as fast as possible, and the one that is as close as possible to the individuals and local communities (both for increasing the detection and for allowing and offering more democratic reactions).

Or rather, the least traumatic among possible roads to environmental sustainability is the one in which every individual (*e. g.* consumers and users) and collective (*e. g.* companies) routinely participate, choosing the solutions favourable to the environment, because according to the stimuli received it seems the best option and most coherent with their values, interests and rationality.

The inspirational principle for policies that try to change people and communities more responsive to the feedback should be bringing to attention the effects of their choices and activities. In turn, the attention could be directed at concrete and direct methods (*e. g.* concrete routine reports on the quantity of disposed products and waste) or at economic methods (*e. g.* the taxation of the use of certain natural resources). Or awareness could be gained through informative activities (with phenomena that are not measurable directly or monetarily, as in the case of biodiversity).

3.7.2 Amplifying the Feedback

At least in theory, the public institutions have many instruments to raise awareness and responsiveness to environmental feedback (just as they have instruments to lower it). Among them, the economic means are most immediate: applying an adequate additional cost to ecological factors already provides a feedback on its own about the ecological implications of one's choices and activities.

The instruments that intervene in the organising of the territory of productive and consumptive processes could be much more efficient. In fact, shortening the circuits, bringing closer different phases of production and consumption, gains the attention of every member of a community with regard to the environmental implications of their choices and activities.

Finally, environmental policies can be instigated that are orientated towards products and tend towards integrating product functions with the feedback, *i. e.* promoting a new generation of products capable of informing the users about the side-effects of their engagement.

Let us stop for a moment on the last option. The easiest step, which is already partly underway, is aimed at informing the user correctly at the moment of purchase about the product's ecological profile. To this belong all the informative instruments we have spoken about (*eco and energy labels*). The problems lay in the design (how to communicate clearly and efficiently such a complex topic) and the cultural background (how to influence buyers to be interested in these indications and actually take them into consideration).

For products that consume resources during their utilisation, indicators could be designed that in real time inform the user about the working status, the need for maintenance, the downgrading and consumption flows, and finally would indicate the appropriate moment for substitution (as in the case of batteries that are equipped with power indicators).

3.7.3 Supporting the Offer of Alternative Solutions

Responsiveness to environmental feedback becomes totally useless (and frustrating) if no alternatives are available. Practical experience and several pieces of social research have shown that often the social sensitivity about environmental issues (which could be seen as a feedback indicator) does not cause an effective change in overall behaviour. Even if it is true that it can be understood as a sign of cultural and behavioural inertia, it is also true and proven that often it is effectuated by the lack of real and useful alternatives. That in many European countries the citizens started with relative efficiency at collecting and separating their refuse was mainly related to the availability of differentiated refuse management systems. *Vice versa*, the waywardness of the same citizens in reducing their use of private transportation, despite being conscious about the related ecological ramifications, is caused mainly by the lack of credible and plausible alternatives (because public transport does not offer equivalent services or because these services are for other reasons inadequate).

The availability of alternatives is obviously one of the crucial issues and the public sector can help here a great deal: these alternatives (*i. e.* a new generation of intrinsically more eco-efficient products and services) and their success (*i. e.* that they are practical) will determine the timing and methods of transition towards sustainability.

Corresponding to its importance, this topic has to be dealt with using all the technological, design and marketing instruments available. The success of alternative solutions depends in turn on how adequately they are proposed; on their economic, social and cultural sufficiency (in this context their ecological motivation, ecological benefits caused by their application, do not count as a separate motivation, but are merged with the cultural and social sphere in integration with other aspects. As when one consumes eco-food, one does it because of the environment, because of one's own idea of well-being and because one wants to appear to be part of a certain social group).

3.7.4 Promoting Adequate Communication

In the end, we can be aware that the feedback is operative and existing alternatives practical, but individuals and communities are still incapable of listening to the former and of practicing the latter.

The third fundamental condition for transition towards sustainability that is based on the collective learning process requires people to have the necessary capabilities. Therefore, the social operator itself has to promote informative activities that further this direction.

This is not about typically circulating the usual information about ecological problems (necessary information, but always under the threat of saturation: at some point people become too tired to hear any more bad news). It is about transmitting precise messages to exactly the right people in order to provide them with instruments that would develop their abilities: to understand the consequences of their actions (to interpret the feedback) and to acknowledge and practice the alternatives, even if these clash with formal behaviour and value criteria (Box 3.3).

Box 3.3 Environmental Policies

With regard to the topics we are discussing, the role of public operator can be seen as consisting of environmental policies and their evolution.

First, ecological interventions were specific and local solutions for urgent cases: they tried to neutralise certain unwanted side-effects of human activities, without amending the causes (*end-of-pipe* approach). Later, this approach was gradually modified and integrated with another way of confronting problems. Therefore, after the first generation of environmental policies were created, others, called all together "second-generation environmental policies", whose objective was already removing the causes of individual problems.

In this way, from dealing with polluting effluent flows, policy moved more and more towards technologies that reduce those polluting effluents (*clean technologies*), through redesigning the products that required such technologies (*clean products*), and finally to reorientation of the demand that motivates the production of such goods (economic and normative measures to influence consumption behaviour towards *environmentally responsible consumption*).

The common goal for all these environmental policies (a goal that is more discussed than practiced) is to reach a point where the demand and supply of products change certain "game rules" and generate a new operational ambient for companies.

3.7.5 Designating Adequate Economical Costs to Natural Resources

The first and most fundamental field in which second generation policies have to operate is economic: no policy of re-orientating the production–consumption system can become successful on a larger scale unless the market is trustworthy. As has happened until now, the different ecological costs are underestimated with regard to other economic costs and therefore do not "tell" that the environment is the actual bottleneck that has to be dealt with. The issue of attributing proper costs to natural resources is a widely discussed argument whose complexity cannot be explained by the short remarks of this book. It just allows us to observe that the issue has evolved from the "*The Polluter Pays*" principle, through the studies of new financial tools, to the projects of reforms embracing the entire fiscal system (*eco-tax reform*), which are supposed to shift the focus from income and work, as it is today, to resource consumption and waste generation (which implies a double effect: it favours creating new jobs and influencing the innovative activities to reduce the weight, and in this case also the economic proportion, of the ecological factor).

3.7.6 Extended Producer Responsibility

The goal of widening the producer's responsibility until the product *end-of-life* treatments has been for years a routine axiom of European and international scenarios. Its official definition by researchers from the University of Lund says that: "Extended Producer Responsibility is an environmental protection strategy to reach an environmental objective of a decreased total environmental impact from a product, by making the manufacturer of the product responsible for the entire life-cycle of the product and especially for the take-back, recycling and final disposal of the product." (University of Lund, 1992).

Its most noteworthy application took place a decade ago in Germany in the form of packaging legislation. Other practices, or proposed practices, include electronic products and appliances, air conditioners and cars (it has to be said that unfortunately this normative, although it has been discussed and in the experimental phase for years, has not found a common method of becoming auspicious and necessary).

The coherence between this normative approach and second-generation environmental policies is based on the fact that Extended Producer Responsibility (EPR) does activate the capacities of the producer to solve this problem (of reducing the weight of refuse). The entrepreneur, who has to take charge of the product *end-of-life* treatment, has in fact an incentive to use his technical abilities and entrepreneurial capacities to re-organise the treatment of disposed products and, most of all, redesign the treatment so that it could be done in most effective way.

From the perspective of the field of action of EPR could be widened to the entire product life cycle, *i. e.* not only to the production and disposal stage, but to proper maintenance throughout the life cycle. Normally, the producer tends to change his role and turn into an operator whose position no longer depends on the sales, but on the results (mobility, entertainment, house cleaning, laundry *etc.*).

3.8 Designers and Co-designers

Let us have a quick look at what could be the role in this process for *design*, referring here to the design community, *i. e.* the sum of all professionals – business- or culture-wise – who belong to this community.

It has been repeated several times that in order to promote sustainable well-being, one has to support the capabilities of people and communities to live better consuming less, *i. e.*, their capabilities to rely less on products and services, and more on contextual qualities and on the possibilities for them to be active.

In the face of this statement, a designer who wants to have a responsible attitude finds himself in an apparently paradoxical situation, at least at first glance: wanting to change the world to a place where well-being is less connected with access to new artefacts (and more with the capacity to improve contextual qualities and personal abilities), but the only way to do it as a designer is to propose new products and services. Is it possible to get past this contradiction?

We believe that the answer has to be yes, provided that the innovative character of the proposals envisioned is not based only on the ecological qualities of the products, but also on the possibilities of triggering and sustaining new concepts of well-being.

Before outlining these topics, it might be useful to lay out some ideas concerning the limits and possible actions of designers inside the bigger interactive system that characterises contemporary society in general and the production–consumption system in particular.

3.8.1 Limits and Opportunities of the Designer's Role

The limits of the designer's activities in general terms are all evident, but because often they seem to be forgotten, it is better to be reminded of them here.

- The designer has neither the legitimacy nor the tools to oblige (via laws) or convince (via moral considerations) anyone to change his way of living. So the designer can only offer solutions, products and services that might be recognised by some as being better than the available offers.
- The designer can only operate within existing social and economical systems answering to the demand proposed by them. This means that it is possible (and obligatory) to be critical about systems in being, but they cannot be thrown aside completely (or they would lose the possibility of having any role whatsoever as a designer).

- In order to point out his opportunities, always in general terms, it can be observed that the designer has to be able to play the following roles:
- Contributing to and enhancing alternative choices, strategies to solve certain problems, that are technically and economically feasible for consumers and users (especially alternatives that are based on a different setting of the problem altogether).
- Furthering the *abilities of users and consumers*, the possibilities of them intervening personally and directly to define the results and choose the means of reaching them (this means to give them the possibility of understanding, doing and also failing, without allowing the mistakes to become irreversible).
- Enhance the imagination of users and consumers (but also producers and public operators), their propensity to *see* the solutions that are still not clearly articulated. This would allow the designer to propose such changes on a cultural level, for value judgements, quality criteria, outlooks on life that would continue to influence the existing system (and finally to orientate the demand for products and services).

3.8.2 *Operative Fields for Design for Sustainability*

What designers can operatively do to further environmental sustainability within the socio-technological systems is based on two complementary levels (that specify, in a similar manner to what has been said about businesses, the strategic intervention scale for designers, as described in first chapter):

- Further the ecological quality of products and services, and in doing so, collaborate with the companies and their development strategies for eco-efficient products and systems.
- Further solutions that allow access to a sustainable life-style, and in this way, collaborate with companies and their development strategies for solving problems that emerge in contemporary society.

The first level of designer's interventions is the main topic of this book and will be covered in the following chapters. It is rather consolidated field, equipped with criteria, methods and quickly adaptable tools for the designing process. Box 3.4 discusses the second level of interventions, which are less consolidated, but crucial against the background of redefining the role of designing.

Box 3.4 New Design Skills

Evolved Industrialisation and Design

It appears that solving problems related to design for sustainability requires skills that are not so different from the skills required to solve problems related to the most advanced fields in the contemporary production system.

It should not be too much of a surprise, as has been said in the second chapter, production today entails not only products, but first of all services, experience and events. This requires establishing new partnerships between a wide variety of social actors. Here it should be added that all this has raised a new demand for designing skills and tools, as for example, *service design*, *interaction design* and *strategic design*.

Until now, these new skills and tools have been neither sufficiently developed nor sufficiently widespread. And first of all, what is most important, their application has mostly been within the production–consumption systems dominant today and has accepted uncritically the current demands, thus accelerating even more the course that would lead to socially and ecologically disastrous results.

Like with so many cases, here the average (and negative) examples also cover other cases, among which many can be highly positive. A more precise observation would actually discover that among more advanced trends currently active in the production system and proposals favouring environmental sustainability exists a mutually fertile partnership, *a positive short circuit* that could give birth to a new form of industrialisation: an *evolved industrialisation* [10], targeted at favouring social innovation with new, unprecedented enabling systems. That is, technological platforms that enforce and favour local initiatives; solutions that facilitate new forms of socialisation and have a lighter ecological footprint. In short, they are sustainable solutions on an industrial scale.

The fact that the designers are working together on a large scale short-circuit of this sort should not be taken for granted. But it is a probable event if design theory and practice will know how to choose the correct direction. In this context it is crucial to be able to stabilise relations between design and various innovative areas.

Design and Innovation

New solutions can be created in different ways: they can be "invented" by designers or be "found" in society. They can emerge from studies of designer evolving together available technologies and social demands. Or they can be the result of selective observations about society and the innovative phenomena they invest in (thus also about new problem-solving methods that can emerge).

If both of the two modalities are viable, the second of the two (using existing examples of sustainable orientated social innovations as a starting point) seems more promising and efficient.

[10] *Evolved industrialisation:* a production, logistic and distribution system that allows different participants to converge towards reaching sustainable goals. To do it efficiently, regarding different reasoning and scale factors, respecting specific contexts, providing importance and leadership for local partners and favouring the active participation of end-users are necessary.

Diffuse Creativity and New Skills

Previous paragraphs have pointed out that in advanced industrial societies certain signals show the existence of groups that are thinking about new methods, new ways of organising, new concepts of well-being (Box 3.1). These signals have been shown to be weak but promising.

All this offers a great opportunity for designer who wants to work in a sustainable direction. In practice, it is all about locating among different weak signals emitted by society those that pertain to new cultural or consumption orientations. And among these the signals that are more coherent with requirements of environmental and social sustainability. Then, it is about how to reinforce these signals, how to reform the proposals to be more concrete and structured, make them more visible, accessible and last longer, to choose and favour one that is more likely to be reproducible on a larger scale.

In this way, a weak signal can turn into an offer of goods and services addressed at a specific sector of demand. After establishing visibility, can this offer appeal for further demand and larger supply, thus widening the market, establishing new quality standards and influencing the more general framework it works in?

So, for the designer it is about favouring the positive interaction between *social innovation* (especially the creative communities and their ground-breaking ideas) and *technological and institutional innovation.* To provide visibility for advantageous cases highlighting their more interesting qualities; to outline existing and future scenarios; to interpret the demand indicated, more or less explicitly, by these cases; to visualise and develop new systems of products, services and experience that would boost their *efficacy* and *accessibility.*

Additionally, performing this role demands from the designer not only to advance his professional communicative, design and strategic skills, but also to further his idea about the place the designer has among other social actors. It appears, in fact, that they have to understand how social actors in current society, as sociology teaches us, "are all designing". How in this society, as previous pages have tried to show, a multiplicity of active minorities are inventing new ways of living and doing.

After locating in wider context what has been just said (and grasping its characteristics), the designers have to recognise that they can no longer have the monopoly over design (everybody designs today); that designing no longer takes place in studios (today one can design anywhere). But they still have their role to play. On the contrary, because in contemporary society, their part, their role as "project professionals", is more important than ever: the designers can enter into the diffuse field of design as moderators, bringing along their specific skills, being able to create "*visions of the possible*" (the ability to envision something that does not exist, but could exist), and to make up strategies to *fulfil goals* (knowing which steps to take to turn probable visions into actual solutions).

This new way of operating in society and the new roles descending from it suggest that, in addition to those already mentioned, designers have to acquire new skills, the abilities to establish partnerships between different social actors (*e. g.* local communities and companies; institutions and research centres), to collaborate with others creating shared scenarios and visions; and to co-design sophisticated systems of products, services and information.

Part II
Design for Environmental Sustainability

Chapter 4
Life Cycle Design

4.1 Introduction

The environmental limits show us clearly that not a single design activity can be carried out without taking into consideration the impact that the product will have on nature. The environmental requirements are now considered necessary from the very first stage of product development onwards, along with the cost, performance, legal, cultural and aesthetic requirements. This is rather beneficial because it is considerably more efficient to work within preventive terms rather than adapt solutions that deal with damage control (*end-of-pipe* solutions). Also, in design terms it is more effective and eco-efficient to intervene with the product instead of dealing with the design and production of other, remedial products for the environmental impact management.

Engagement of an environmentally conscious strategy from the very beginning of the design process would help to prevent or limit the problems, instead of losing time (and health and money) with redressing the damage already done. In this way it is easier to combine both environmental and economic advantages.

During the second half of 1990s a new discipline of *the design of products with low environmental impact* emerged that was more concretely and realistically able to deal with the complexity of its subject: it became clear what is meant by the *environmental requirements* of *industrial products*; the concept of the *Product Life Cycle* (for design and assessment) was introduced and the concept of *functional unity* was re-contextualised in environmental terms. The following paragraphs will examine these concepts.

4.2 Environmental Requirements of Industrial Products

During the 1990s, consequent to certain studies and new assessment methods, it became possible to evaluate the environmental impact of the input and output between the technosphere of a certain product and the geosphere and the biosphere. This allowed us to specify what is meant by *the environmental requirements of industrial products*.

These studies emanated from an obvious (though not easily calculable) realisation: each environmental effect is based on an impact of exchanging substances between *nature* and the production–consumption system. These effects can occur in two directions:

- As *input:* extracting substances from the environment
- As *output:* emitting substances into the environment.

Obviously not all impacts are harmful or equally as damaging (to pour into the environment 1 kg of water is quite different from throwing 1 kg of asbestos powder!).

So, which ecological impacts have to be considered in relation to environmental requirements?

When speaking about input – extracting resources for a certain product – the first harmful effect is the exhaustion of the resources, which will first-hand create a social and economic issue of lack of resources for future generations[1].

Similarly related is the issue of the alteration of the ecosystems' balance, for example, the deforestations due to the use of timber in construction (of various types of artefacts) or in heating systems. Over the course of time, this human action has made the land more vulnerable to erosions, and determined the extinction of some species that inhabit the forests.

Finally, there are the harmful effects of *output* connected to the extraction processes, *e. g.* the oil-leaks during extraction and transportation processes.

The main environmental impacts due to emissions usually connected with different products are:

- Global warming (greenhouse effect)
- Ozone layer depletion
- Eutrophication
- Acidification
- Smog
- Toxic emissions
- Waste

Other impacts could also be added, such as electromagnetic and genetic contamination. But we are not going to describe these effects here any further[2]; just to underline, that when speaking about environmental requirements, design aims to reduce the influence of products on these effects.

The real issue, the actual question is, if it is possible to associate these effects with a product, for example, the amount of global warming effect relating to mobile phones, which are owned by almost all habitants in industrialised countries.

---

[1] *Cf.* definition of sustainable development, Chap. 1.

[2] For further elaboration *cf.* Appendixes.

The answer is yes, to reach such a conclusion, the methodology of *Life Cycle Assessment* (LCA) can be employed, among others, a method of estimating the above-mentioned effects caused by the *input* and *output* of all processes more or less referable to a certain product [3].

There are two key approaches, which have great enough repercussion on design approaches that aim to integrate environmental requirements into industrial products. These approaches, which will be described soon, are:

- The Product Life Cycle
- The functional unit

4.3 Product Life Cycle

4.3.1 Introduction

The concept of life cycle that is under discussion here refers to input–output exchange processes between the environment and the whole set of processes that entail the entire lifetime of any given product [4], meaning that the product is analysed according to its energy, resource and emission flows during its lifetime. Thus, the life cycle encompasses all stages of the product, starting with mining for necessary resources and manufacturing its components until the last end-of-life treatment.

The entire *life* of a product can be described as one set of activities and processes, while every one of them consumes a certain amount of resources and energy, goes through series of transformations and triggers emissions of various kinds.

To help to map out the actual life cycle of a product, these processes are usually divided into the following phases:

- Pre-production
- Production
- Distribution
- Use
- Disposal

The concept of life cycle is able to adapt a system vision over the product's input–output during all phases, analyse and assess its environmental effect, together with economic [5] and social [6] influences.

---

[3] *Cf.* definition of LCA in Chap. 12.

[4] The term *Product Life Cycle* can cause some misunderstanding, due to its usage in a management environment, where it refers to a product's entrance, stay and exit from the market.

[5] Life Cycle Costing.

[6] The social impact of life cycle is currently being studied by various authors.

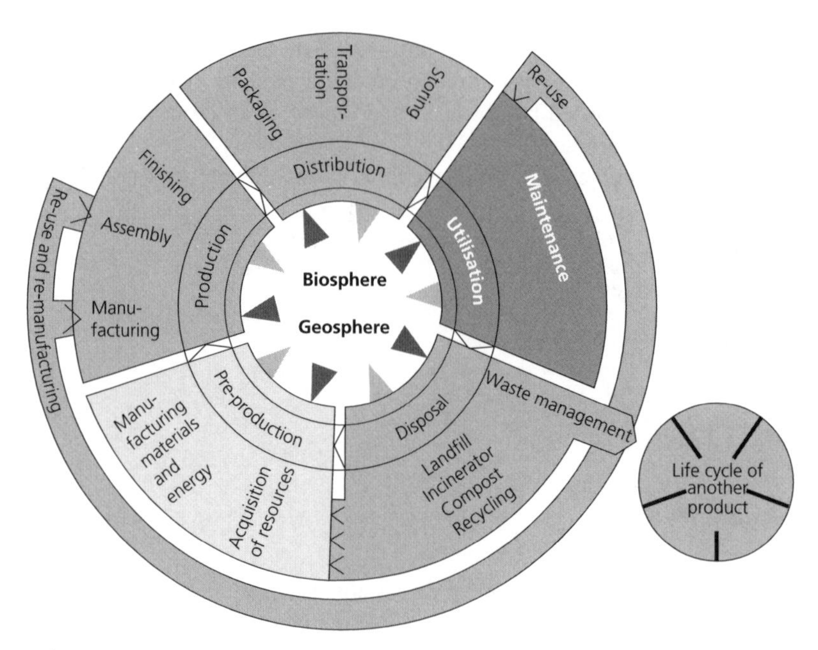

Fig. 4.1 Product-system life-cycle

Figure 4.1 visualises the system approach to the set of possible physical and chemical relationships between the production system during its different phases and the biosphere [7] and geosphere [8].

In the following paragraphs these phases will be outlined and clarified, along with the corresponding fundamental processes [9].

4.3.2 Pre-production

In the pre-production stage necessary resources and semi-finished products are prepared as components of the manufacturing of the final product. The following sub-stages can be briefly summarised as:

- Acquisition of resources
- Their delivery to the production area
- Their transformation into raw materials or energy

[7] The biosphere is the total sum of living organisms, or more precisely, the part of the global territory that provides the vital means of subsistence for organic life.

[8] The geosphere is the total sum of water and land masses.

[9] These topics are amplified in the chapters on the respective designing strategies.

Raw materials and energy are produced from:

- Primary or virgin resources
- Secondary or recycled resources

Primary resources come straight from the geosphere and can be classified as:

- Renewable primary resources
- Non-renewable primary resources

Non-renewable resources are mined from the earth, while renewable resources, from biomass, are normally cultivated and harvested. In both cases raw materials go through a series of treatments [10].

Secondary raw materials are derived from scraps and discards of the production–consumption system. More precisely, these materials can be reprocessed at two stages:

- Pre-consumption
- Post-consumption

Pre-consumer resources consist of waste and refuse discharged during the production processes. While post-consumer resources are acquired from goods and packages after they have passed the end-user. These resources, especially post-consumer ones, have to be reprocessed before their re-utilisation in new products is rendered possible [11].

4.3.3 Production

We can roughly distinguish three fundamental stages in production:

- Processing of materials
- Assembly
- Completion

Delivered raw materials are stored in a production area and at the proper time conveyed to the machinery and processed into components, which afterwards will be assembled until the completion of the final product. At this point, other finishing processes can be applied, for example, painting and polishing.

Most of the goods demand a great variety of raw materials for completion, either directly or indirectly. Direct materials will be contained, after transformation, in the final product; meanwhile, indirect materials are combined in machinery or facilities used for production. Other activities and processes applied during this stage could be research, development and inspection, and management of these activities.

---

[10] For example, the processes needed to transform iron ore into ingot or sheet steel; bauxite into extruded aluminium; refined oil into monomers and then into plastic granules.

[11] This matter belongs to the topic of *secondary raw materials*; its characteristics and problems are delved into more thoroughly in Box 8.1 in Chap. 8.

4.3.4 Distribution

Three fundamental stages that characterise distribution are:

- Packaging
- Transportation
- Storing

The final product is packaged in order to reach the end-user intact and functional; shipping is carried out by different means of transport (train, lorry, ship, plane, pipeline *etc.*), either to an intermediate station or directly to the sphere of application. This stage entails not only energy consumption for transportation, but also resource consumption to produce the means of transport as well as storage facilities. In reality cases can occur where the line between distribution and production[12] blurs. However, it is important that the periods before or after these operations are taken into consideration.

4.3.5 Use

Two fundamental stages that characterise utilisation are:

- Utilisation or consumption
- Service

Thus, goods are for a certain period utilised or, according to their characteristics, consumed[13].

The utilisation of goods in most cases consumes resources and energy and leaves behind refuse and waste. Besides, they can face maintenance[14] and servicing; repair of damages or damaged parts; or replacing[15] obsolete parts.

Products remain in utilisation until the consumer does not decide to get rid of them, or, to express this more correctly, does not dispose them; at a certain point this can happen for various reasons[16].

4.3.6 Disposal

At the moment of product disposal a series of options for its consecutive destiny opens up:

---

[12] For example, concrete is processed in the truck while being transported to the building yard.

[13] Articles of food are consumed; meanwhile, the TV set is used for a certain amount of time.

[14] *Maintenance* comprises the activities of periodic prevention and light adjustment.

[15] *Upgrading* adjusts semi-durable products, substituting parts that have become obsolete.

[16] These arguments are also dealt with in Chap. 7.

- It is possible to restore the functionality of the product or its parts.
- It is possible to recover the component materials and energy of the disposed product.
- After all that, it is possible to regain nothing.

In the first case, the product or its parts can be *reused* within the same or some other function. A reusable product has to be recollected and delivered. Otherwise, the product can be *re-manufactured*[17] and go through phases that allow it to be used as new.

In the second case, component materials can be *recycled, composted or incinerated.* There are two different courses for recycling:

- Open loop recycling
- Closed loop recycling

Closed loop recycling implies a system where recycled materials are used, instead of primary raw materials, inside the same production system. That is, they are used for manufacturing similar products or components to those from which they were disposed of.

Meanwhile in *open loop recycling* materials are directed to a production system that differs from their original one[18].

In any case – in open or closed loop – recycling always entails a series of processes and stages that comprise recollection, transportation and pre-production of secondary raw materials[19]. Also, incinerated materials have to be first collected and transported.

Otherwise unused products are dumped into (more or less officially) landfills or just dispersed into the environment. In the case of the legal landfills disposed products still have to be collected and delivered first or, in the case of toxic or harmful substances, even go through preliminary treatments[20].

4.4 Additional Life Cycles

As it appears, the concept of *life cycle* takes into account every stage related to a given product, from pre-production to disposal. But, more precisely, the defini-

[17] *Re-manufacturing* is an industrial process whereby worn-out products are partly processed to return their characteristics to a similar level to what they were initially.
[18] Occurs only with post-consumer materials.
[19] For the logical rigour, in reality these activities, resource consumption and generated emissions, should be counted in the pre-production stage of the product that is going to exploit the recycled materials. In fact, if the transportation and extraction of primary raw materials are part of it, so should the secondary raw materials that are extracted from the scraps and discards and later transported to the re-manufacturing sites.
[20] Illegal landfills present both environmental and social dangers; in fact, there exists a market for toxic waste run by organised crime.

tion of the sum of processes (and of the correlated input–output) refers to the function delivered by the product (as we are going to see in the next paragraph).

This means that, in the case of many products, along with the actual life cycle, the life cycles of other products have to be considered that are functional with regard to the services offered to the user[21].

Packaging and various consumable products are clear examples of this, but are, nonetheless, subsidiary systems that guarantee a product's functionality. Like all other goods they also have their corresponding input–output exchanges with nature that induce environmental impact.

The packaging is an *additional*, quite important life cycle being adopted in most products nowadays. It is in every aspect a product on its own, and accordingly has its own life cycle: pre-production, production, distribution, use and disposal. The functions of packaging – to contain, protect, transport and inform – become operational after they come into contact with the corresponding products, meaning that the product distribution stage coincides with the package usage stage.

4.5 Functional Approach

The second criterion in assessing the environmental impact of a product is the *functional unit*. According to this approach, it is not the product that is under assessment, but the impact of the set of the processes employed to *satisfy* a certain function.

Before the exact definition, let us take an example, a car and a bus. When comparing the life cycles of these products, even without complex calculations, it appears that the bus has a greater impact in every phase: it needs a greater amount of resources to be pre-produced, greater processes while being manufactured, heavier transportation, it consumes more resources while being used and leaves behind a greater amount of materials for disposal. When the impact is considered according to its function, which in this case could be the transportation of one person per n kilometres. Let us assume for simplicity that one car carries two persons on average and the bus carries 20. Now, in relation to their supplied function, it appears that we should have compared the bus with 10 cars instead and that the impact of the bus would have been far smaller than that of the car.

The *functional unit* is a performance of the product that is being assessed; the function or service supplied by the products has to be studied, not the physical product itself.

---

[21] Taking for example coffee machine, with the function to make good coffee. Without descending into the reasons behind our desire to drink coffee, but instead underlining the possibility to drink it, not only because the coffee machine can make it (in accordance with used energy), but equally due to access to water and coffee. Now we have to calculate the *input–output* processes of coffee and water as well as the disposable coffee filters. In other words, their production, transportation and disposal have to be included accordingly to the life cycle of coffee machine.

If one would like to compare the product before and after (re)designing, one would have to match products, services and processes that are functionally equivalent. Because of this the *functional unit* always has to be defined to become the measure for comparisons.

The second fundamental criterion is therefore based on designing the function that the product has to supply rather than the product. Because of this association the overall analysis reckons whether the environmental impact has been minimised or reduced and how much.

As mentioned, it is a basic concept of design for sustainability, which means that design-wise it is a step back from the product, starting with its *function*, with the *satisfaction* it is supposed to bring to the user.

Here, the *function,* the fundamental topic for design culture (once a guideline, but criticised by many), gains a new meaning and vitality confronted with environmental issues[22].

4.6 Life Cycle Design

The design discipline that deals with the environmental requirements of industrial products in the aforementioned way is called *Life Cycle Design* (LCD). This expression is closely related to *ecodesign* and *design for environment*, but LCD is more appropriate.

Besides the terminological questions, it establishes and stipulates a new design approach that entails greater vision than is customary. Together with the key criteria described in the previous paragraphs, the LCD approach means:

- *Wider design horizon*: from product design to LCD, designing stages of the life cycle.
- From product design to design of function, designing the *satisfaction* of the product, *e. g.* from car to mobility.

So the goods have to be designed within all their life cycle stages. All activities needed for manufacturing its materials and then the product itself, its distribution, use and finally disposal are all considered to be one totality. Thus, it takes from product design and goes to product system design, all the events that determine and accompany a product during its lifespan.

A design with a systematic approach allows all consequences of a designed product to be identified, including at stages that are not traditionally included in design process. The future responsibility of product development will be to design product life cycles. In this way, it is possible to identify and efficiently engage the objectives of reducing environmental impact.

Evidently it is necessary to integrate an environmental perspective into all the aspects of development processes, not only in design, but also in management and marketing.

---

[22] Tomas Maldonado speaks of "environmental neo-functionalism", 1998.

4.7 Life Cycle Design Objectives

Discussion of LCD does not mean to focus only on environmental requirements; it is intended to be a more general approach to design. Nevertheless, when we do consider, as it is in our interests, the environmental requirements, then the objectives will be to minimise the burden on the environment associated with the product during its life cycle and related to its functional unit.

In other words, the goal is to create a systemic idea of the product to minimise the input of raw materials and energy, let alone the impact of all emissions and waste, both quantitatively and qualitatively, calculating the harm of all effects[23].

The greater vision leads the designing process to take into account all activities during its life cycle and relate them to the set of *exchanges* (the *input and output* of various processes) they have with nature. For this must be defined, within the design process: a profile of the stages of the product life cycle, starting with extraction of raw materials until the disposal of waste and residues.

4.8 Implications of Life Cycle Design

To bring out the advantages of Life Cycle Design, let us start first of all with a statement that many products have been designed where environmental impact has been greatly reduced at one stage, but the overall impact has increased, because it has worsened (at another stage and in relation to the functional unit) more than it has improved.

A glaring example is cardboard furniture, which contrary to the general view is a disaster in environmental terms. It is true that the pre-production and production of a cardboard seat is less demanding than those of other seats made of more traditional (and durable) materials. It is just that cardboard items become unusable very quickly and have to be substituted, and substituted again and again; meanwhile the traditional seat is still in use. Thus, in comparison with the functional unit, as we have learned, the overall impact of sitting on a cardboard seat will be exponentially higher than that from a seat made of other, more durable materials.

Here, the LCD approach, which takes into account all stages of the function, has an advantage in identifying the priorities of the product design concerned, priorities that differ from product to product.

The disadvantage, sort of, remains in the more complex design process (but not more complicated, in fact for many reasons, even more stimulating).

The first reason behind the complexity is the required amount of information about the *input–output* of processes and about their impact on nature. We

[23] Some substances may have close to no effect, or at least a very small effect, even in large quantities; thus, they should not provoke any concern; others, meanwhile, as in the case of toxic substances, can effectuate grave results, even in small amounts.

will see that for this end there exist various databases, methods and software tools[24].

Another reason, regarding mainly the disposal, is the unpredictable techno-economical evolution. Not everything is known during the design process. Technological, normative and cultural contexts are in continuous evolution and hard to predict with any certainty (especially for the disposal stage, but also for usage): how exactly will the system's conditions be established?[25].

So we can design for example a (durable) product to facilitate the recyclability of materials, but in 10 years' time, when it is finally recycled, the technology of recycling will have changed compared with what we foresee today, and might not be adequate for our product or its materials.

The same story applies for the operational technologies of the same product, which could become more efficient over time.

In the end, a designer/producer is rarely the only one responsible for the whole product system; at times other participants take over or control the various stages along the life cycle: raw material and semi-finished product suppliers, manufacturers, distributors, users, companies and organisations dealing with disposal.

This is why it often remains difficult to design and later remain responsible for the entire product life cycle system[26]; the degree that could, during the design process, effectively determine the set of processes that would accompany the product throughout its life cycle varies greatly.

4.9 The Design Approach

However, these observations should not hold back the transition to the *Life Cycle Design* approach, but rather should be taken as adjustments to an application's modality.

In the end, an efficient and accurate LCD approach should try to minimise environmental impact at all stages, but also according to the best and most probable system configuration available. Which conditions (and therefore adjustment) depends on the type of product and the dimension in which companies work.

After all, it is possible to design with the objective of minimising environmental impact within the production system controlled by the producer (designer) or when he has only partial control.

Summarily, the following cases can be identified:

- To design knowing that the responsibility/control lasts for the whole life cycle. For example, a manufacturer, who has to deal with the disposal from his production process in compliance with normatives regarding extended producer responsibility.

---

[24] *Cf.* Chap. 12.

[25] *Cf.* Sect. 4.13.

[26] *E.g.* it rarely happens that the same manufacturer is the same actor who takes care of disposal.

- More often product design happens to become part of an system that already exists, totally or partly controlled by other *actors* (a system where others are responsible/control certain stages). So a young designer who wants to optimise the disposal service in his hometown first has to know how exactly the entity in charge of waste management has organised the refuse collection and recycling.
- Or rather can a designer be in a partnership with other *actors* and "co-design" for a certain stage of the life cycle only? For example, a manufacturer draws up a contract with a producer who is active in waste disposal to retrieve materials within a certain territory in which products were sold.
- In the end, because the results of the technological and economic evolution are hard to foresee, it is necessary to design flexible solutions, to enable their adaptation to those changes.

A designer can engage the LCD approach in all these cases and succeed with fewer difficulties in identifying the environmental impacts of certain products and to reduce them efficiently, without limiting himself to sticking to only one or other stage of its life cycle.

4.10 Strategies of Life Cycle Design

The following chapters will present the strategies, guidelines and design options involved in integrating environmental requirements in product (and service) development.

Designing a good product obviously cannot be based solely on fulfilling environmental requirements, just being an eco-efficient product[27]. The discourse of this book takes for granted the necessity to satisfy other typical requirements of product design: performance, technological, economic, legislative, cultural and aesthetic requirements.

Strategies within this radius are the following:

- Minimise material and energy consumption.
- Select low impact processes and resources; select more eco-compatible materials, processes and energy sources.
- Optimise product lifespan; design durable and intensely usable products.
- Extend material lifespan; design with the purpose of increasing the value of disposed materials via recycling, composting or incineration.
- Facilitate disassembly; design with the purpose of separating the parts or materials.

---

[27] Eco-efficiency is a term proposed by the World Business Council for Sustainable Development (WBCSD), and refers to the relation between the value of the product (supplied satisfaction) and its environmental impact (pollution and resource consumption); it indicates the degree to which the reduction in the environmental impact during production, distribution, use and disposal, and the increase in the quality of the goods and services are joined.

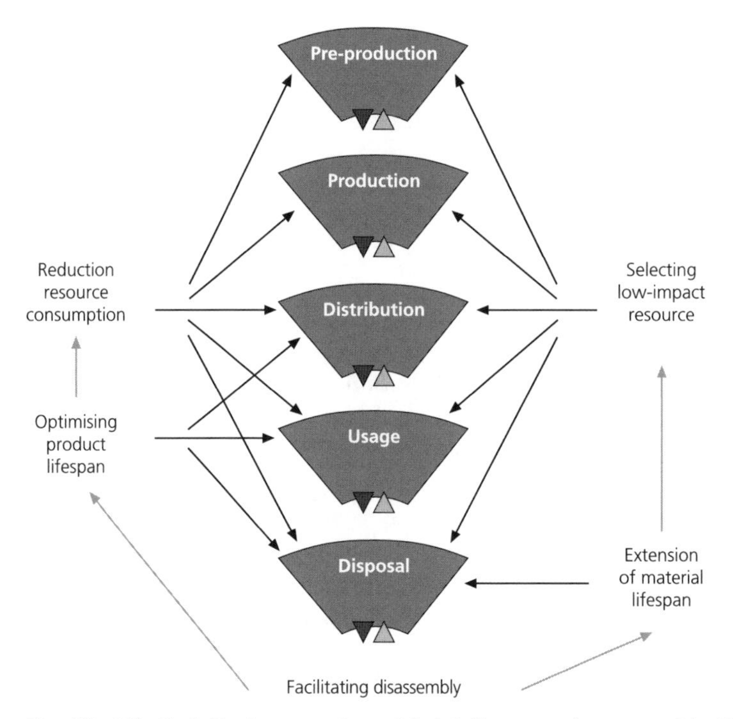

Fig. 4.2 Life Cycle Design strategies and their influence on the stages of the life cycle

Reducing consumption and selecting low-impact resources are the objectives for all stages of the life cycle.

Optimising product lifespan determines an overall reduction in the environmental impact during the stages of pre-production, production, distribution and disposal (by means of the reduction of product flow necessary to satisfy particular needs and desires); the impact during the stage of usage is more of a socio-cultural phenomenon, since this requests, to a different extent, changes in the ways products and services are utilised.

Extension of material lifespan is an objective for the stages of pre-production (less raw material consumption) and disposal. Facilitating disassembly is needed for optimising product lifespan and extending the materials' lifespan (Fig. 4.2).

4.11 Interrelations Between the Strategies

When a product becomes more durable than before, on the one hand it reduces the by-product waste and on the other it indirectly avoids further resource consumption for the production and distribution of new products needed to substitute short-term goods.

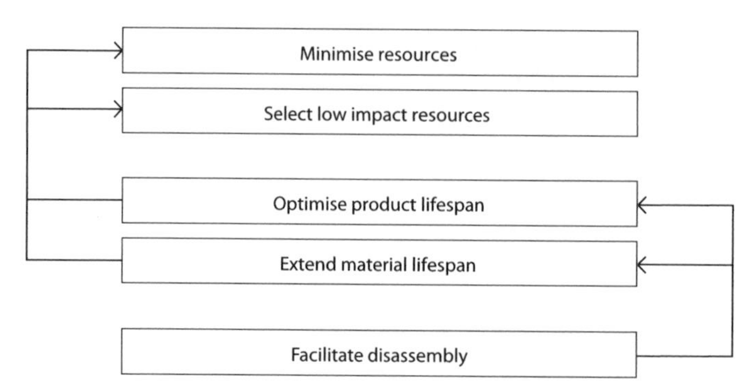

Fig. 4.3 Functional hierarchy of Life Cycle Design strategies

Besides, the extension of the material lifespan (due to recycling, composting or combusting) decreases the discharge volumes and raw material consumption of new, similar materials.

Finally, it should be introduced here (and soon explained) that *facilitating disassembly* is a strategy useful for both *optimising product lifespan* and *extending material lifespan*; therefore, also aiding the *reduction of resource consumption* and the *selection of low impact processes.*

Actually, both optimisation of the product lifespan and extension of the material lifespan will consequently bring along a selection of low-impact resources and processes. And facilitating disassembly is needed for optimising product lifespan and extending the materials' lifespan (Fig. 4.3).

Every strategy outlined can follow (as will be shown shortly) certain *guidelines* and specific design *options.*

These strategies on their own seem to be the very same objectives as those of a design project. Actually, to be efficient, they have to be used only according to the objectives and priority requirements of a given product (to be designed).

4.12 Priorities Among the Strategies

Beneath every life cycle design, lies an environmental objective of reducing to the minimum both the *input* of materials and energy and the impact of any emission or waste, *i. e.* the *output*, of the production system.

It is presumed that for every product, whether it is a car, a table, or something completely different, some strategies have greater priorities than others, such as for a table it would be more important to optimise its lifespan rather than reduce its resource consumption, because unlike a car, a table does not consume resources while in use.

In the end, one strategy could also conflict with another. For example, using biodegradable materials is appealing, but for many products durability is more important and a biodegradable material could compromise that.

Therefore, before even starting to design, it is important to identify the priorities of the strategies, correspondingly for each product and its function, to graduate the importance of the available scenarios.

For a context in which a life cycle depends more on the duration of a product and on the repeated use of its components and materials, it would be more efficient to start with strategies of *product lifespan optimisation* and *material life extension*[28].

On the other hand, the *product lifespan optimisation* and *material life extension* are nothing but indirect ways of *minimising resource consumption* and *selecting low impact resources*.

In operative terms is rather improbable that employing a single strategy would be the best way to satisfy all the environmental requirements of a given product. Therefore, is better to create a set of environmental strategies and design choices. The presence of such a set would be helpful to avoid starting off with low-profile propositions when better development options are available. If various strategies are simultaneously employed then at best they can be synergic, though they could also clash with each other[29].

Thus, it has to be understood how to decide whether simultaneously adapted strategies create more trouble than advantages, to set the priorities corresponding to the objectives and decide which road to take and in which modality.

On the other hand, it is necessary to understand the product and its system characteristics in order to efficiently pose the objectives and environmental requirements (Box 4.1). It is the only correct way to estimate possible improvements for environmental impact in relation to the entire life cycle.

For analysing and supporting design decisions several methods and tools have been developed and are in constant evolution[30].

Besides the conflicts between strategies, the environmental scenarios can also clash with the requirements of traditional design practices, whether they deal with performance, costs, normatives, cultural or aesthetic characteristics. For example, an extended product lifespan (to lessen the environmental impact) can be seen as a possibility of reduced sales. An example of common grounds instead is the reduction of energy consumption during the production stage[31].

From the sustainability perspective, the environmental requirements are of course of primary importance, but it is also true that for a realistic solution to follow the criteria of environmental impact reduction, it must be economically feasible and socio-culturally attractive.

---

[28] It is noteworthy that these strategies, more than the two previous ones, are contradictory in respect of current industrial and consumption tendencies in contemporary society and advanced industry. Besides some interesting cases outside the mainstream, the dominant view prefers short term products (particularly throw-away goods) and diminished producer responsibility for end-of-life treatments. The restraints and economic opportunities will be discussed further in Chap. 7.

[29] *E.g.* using recycled polymers (for less impact) will clash with the guideline to reduce the overall weight of a product. In fact, in order for a recycled component to have equal endurance, it has to be heavier.

[30] *Cf.* Part III.

[31] *Cf.* Chap. 7.

Therefore, the employed strategies have to satisfy the entire set of requirements. If the performance of a product is compromised for the sake of environmental improvements, the benefits of such design can only be illusory.

Box 4.1 Product Categories

To engage environmental strategies efficiently it is important to identify the category of the product under examination. Here is a schematic syllabus that could prove to be useful for such cases[32].

Consumer Goods (throw-away goods). With two significant subdivisions.

- Goods that become used up, like food or washing powder. Obviously, there is little sense in designing them to be more durable. But it is important to concentrate on minimising their resource consumption and to select low impact resources.
- Throw-away goods that can be reused, recycled or substituted, such as packaging, newspapers and throw-away razorblades. With products with greater impact during the production and disposal stage it could be fruitful to extend their lifespan by substituting with other reusable goods[33] or by making them reusable (at least partly[34]).

Durable Goods (reusable goods). With some significant subdivisions.

- Goods that consume little or no resources while in usage[35]. Their impact occurs mainly during the (pre-)production, distribution and disposal stages. Thus, it is of primary importance to minimise the resource impact and consumption of production and distribution activities. The impact of disposal can be minimised by extending the material lifespan, but frequently it is more efficient to extend the product lifespan, especially in cases of cultural obsolescence.

Goods that consume resources and energy during use and maintenance. For this category the extension of lifespan might be questionable. In this category there are more important other strategies, particularly reduction of their resource consumption. In fact, extension of lifespan might be counter-productive, especially in cases in which technological evolution may create opportunities for more efficient[36] products (using less energy and resources besides causing fewer emissions) in the near future.

---

[32] *Cf.* Heskinen (1996).

[33] *E.g.* substitute throw-away packaging with reusable packaging.

[34] *E.g.* substitute one-piece toothbrushes with ones that have a replaceable bristle brush.

[35] *E.g.* furniture or bicycles.

[36] *E.g.* greater energy efficiency of washing machines that has been raised in recent years by about 40–50% or the *standby* and *timer* functions of the electronic appliances.

The importance of product lifespan can become paramount for products with a fast obsolescence rate (technological or cultural). Some of their parts could be substituted[37] effectively to upgrade their efficiency or to downscale the activities of production and disposal only to those needed for the substituted parts[38].

4.13 Design for Disposal

Special attention has to be paid to the end-of-life strategies. Not because this stage has extra high environmental impact, but because at this stage the manufacturers and designers are less involved than at any other. Above all, the period between design and disposal sets its own limits. Here, recyclers, producers and designers have to overcome several problems related to the time spanning from the moment a product is designed and the moment it is disposed of (for reuse, reconstruction, recycle, incineration and discharge).

On the other hand, technologies and treatment (recycling, combustion, discharge) costs evolve with regard to what was known during the designing of the product. This inevitable fact adds its own uncertainty and requires solutions that are both flexible and re-adoptable. This is true today and will be true in the future.

In particular, three stages of graduated intervention and ecological reorientation can be outlined:

- Immediate – corresponding to products already designed and ready to be discarded. It is no longer possible to change the production characteristics[39], only the modifications available to improve the processes of treatment, retrieval and valorisation of components and materials.
- Short term – corresponding to products at the design stage and with a short-term lifespan. Modifications available are incremental[40] because of production system inertia, which would not allow immediate redesign of the entire product.
- Medium to long term – corresponding to products accessible for a complete make-over and with a medium to long lifespan. Radical innovations[41] can be made within the product and with regard to its end-of-life treatment.

---

[37] *E.g.* construction materials and tyres.
[38] *E.g.* updating computer hardware.
[39] Designed from 5–15 years ago, thus without any strategy to ease their disposal.
[40] *Incremental modifications* exclude changes in material flows or production organisation.
[41] *Radical innovations* that entail a considerable make-over of the production system.

4.14 Environmental Priorities and Disposal Costs

In environmental terms it is preferable to reuse the product or its components, rather than to recycle or incinerate its materials (not to mention the possibility of dumping them in a landfill site)[42]. Nowadays, unfortunately, the high costs of maintenance, reparation, reuse and re-manufacturing (today mainly labour costs) predispose to recycling and incineration. Actually, their costs have also been rising recently; nevertheless, they are preferable to dumping in landfill sites (Fig. 4.4).

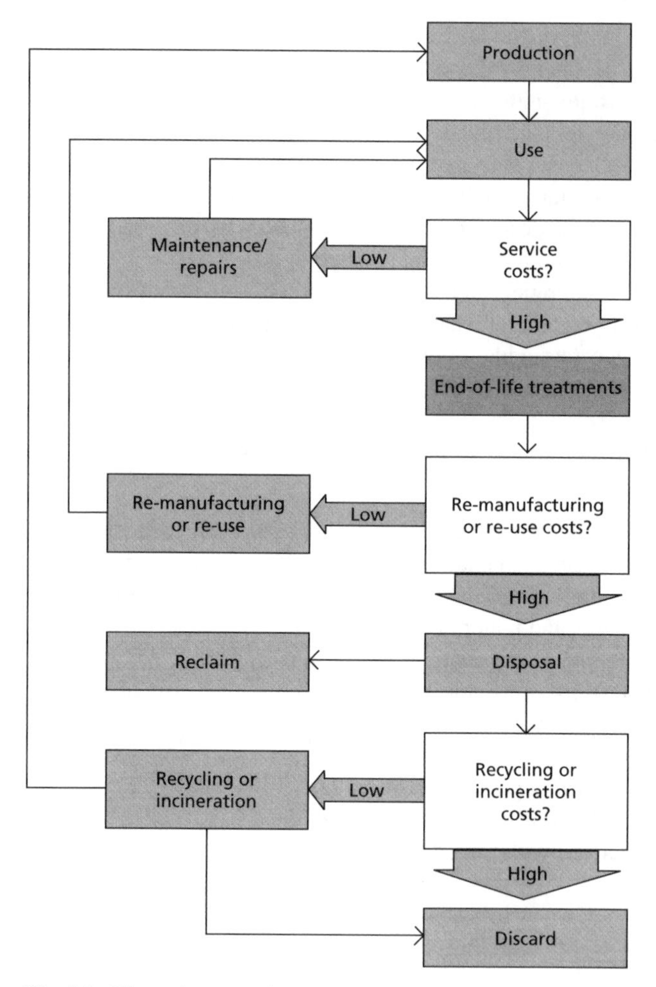

Fig. 4.4 Disposal costs and prospects

[42] Of course this is a generalisation, while for certain products other environmental priorities can exist. Throw-away syringes, for example, are preferable in order to minimise the circulation of viral diseases.

4.15 Current State of Life Cycle Design

Nowadays, *Life Cycle Design* or *ecodesign* of the product has formed a structured discipline, has been provided with a consistent theory, clear designing guidelines[43], methods and tools, as well as study courses in universities.

Life Cycle Design is a lot less advanced in design practice for economic reasons such as the fragmented structure of actors and their economic interests during a life cycle[44]. On top of the industrial inertia towards re-orientation of the production system caused by increased operational and management costs, there is short-sightedness in respect of the long-term perspective.

Other problems arise from a lack of generalised consensus within the science and design community on several aspects of estimating the sustainability of a given product or service that has been achieved. Up to today, various criteria for estimating the environmental impacts are in circulation, and their results can vary within one of the criteria themselves[45].

---

[43] It is generally accepted (irrespective of the subtleties and stresses of different authors), that the environmental requirements of a product entail the following criteria: designing to reduce the amount of materials and energy necessary at any stage of the life cycle; choosing non-toxic and unharmful, materials, processes and energy sources with the greatest renewability and the smallest exhaustibility possible; to design and esteem recycling, incineration or composting the discarded materials; designing for the disassembly/separation of the materials and components.

[44] *Cf.* Chap. 10.

[45] *Cf.* Part III for criteria and methodologies of assessing environmental impact, together with tools that are meant to integrate the environmental requirements during the development of a product.

Chapter 5
Minimising Resource Consumption

5.1 Introduction

Minimising resources means reducing material and energy consumption of a certain product; it is better if it is at every life cycle stage and of the entire service offered by the product; that is, corresponding to the *functional unit*.

This strategy concerns a quantitative saving for the environment and in sustainability terms has two implications. First, with regard to the input flow, it helps to save natural resources for future generations[1]. On the other hand, with regard to the output into nature, the reduction of resource consumption, which is trivial really, overrides the environmental impact of what is not consumed.

Using up fewer materials diminishes the impact not only because fewer materials are extracted, but also because it avoids fewer processing, transportation and disposal costs. Analogically, smaller energy consumption diminishes the impact otherwise caused by its production and transportation[2].

Materials and energy have their economic costs as well as their ecological ones; thus, minimising their consumption is an overall source of saving, though currently consumption reduction during use is not necessarily among the goals of companies and therefore it can not be claimed that it translates into highly efficient product design.

It is especially true in a sales-based economy, but, as will be demonstrated in the following paragraphs[3], it is also possible to develop a more integrated product service system in which the user does not necessarily own the product. This new form of supply carries some interesting intrinsic values with regard to eco-efficiency.

---

[1] *Cf.* definition of sustainable development in Chap. 1.

[2] Methane is an excellent energy source for heating and domestic consumption, but losses occurring during transportation are highly polluting; this gas more than any other is the cause of global warming.

[3] *Cf.* Chap. 10.

In the interests of clear understanding and an efficient supporting structure for designers, the guidelines are divided according to the nature of the resource:

- Minimising materials consumption
- Minimising energy consumption

5.2 Minimising Material Consumption

This section illustrates the guidelines for minimising not only the materials used for and by the product, but also the scrap and discards created during the production of goods and their packaging, including material consumption during the design process.

In particular the following guidelines will be illustrated:

- Minimising material content
- Minimising scraps and discards
- Minimising or avoiding packaging
- Minimising materials consumption during usage
- Minimising materials consumption during the product development phase

5.2.1 Minimising Material Content

Minimising the material content is considered, as said above, to be in compliance with the function employed by the product. Without going into discussing what is actually functional and what is not, it should be noted that it is important to define the functions critically, regarding them in the social, cultural, and economical context in which the saving will assume a new and valuable position.

The minimisation starts from thinning down the sides of the component (together with employing certain geometrical forms that would preserve the necessary rigidity) and finishes with the actual *dematerialisation*, *e. g.* if it is possible to substitute the *hardware* for *software.*

Miniaturisation is another interesting tendency, since technological development has allowed, especially in the electronics field (microelectronics), to reduce drastically the amount of necessary materials required by a certain function.

An additional analysis has to be undertaken whenever the constituent materials are substituted by another[4], *e. g.* the thickness optimisation has to be recalculated.

---

[4] As in the case of exchanging metal parts for plastic ones.

Finally, if a product aggregates the functions of more than one product, it should be compared with the total amount of materials required otherwise. This is why products used more intensively or more extensively have low material intensity. But this will be discussed elsewhere[5].

Guidelines for material content minimisation:

- *Dematerialise* the product or some of its components (Examples 5.1)
- Digitalise the product or some of its components (Examples 5.2)
- Miniaturise
- Avoid over-sized dimensions (Example 5.3)
- Reduce thickness (Example 5.4)
- Apply ribbed structures to increase structural stiffness
- Avoid extra components with little functionality

Examples

5.1.1 "Ikea air" is a set of onsite-assembled sofas and armchairs, designed by Jan Dranger, models of seats furnished with inflatable air cells made of plastic (PE) and an external cover. Cells can be inflated with a normal hair dryer, and need not to be re-inflated for 3 years. The inflatable furniture minimises the use of resources dematerialising the product: the quantity of material used is, on average, 15% of that required for a conventional armchair/sofa.

Example 5.1.1 "Ikea air" series

[5] *Cf.* Chap. 7.

5.1.2 All elevators by Kone MonoSpace use the Kone EcoDisc, an axial synchronous engine on permanent magnets (that works at an adjustable frequency and without a reducer) without an engine room, which is considerable dematerialisation, in addition to the notable reduction in the space occupied. Also, the elevators are equipped with highly efficient LED lights, along with alternative energy sources consisting of photovoltaics and halogen-free cables.

Example 5.1.2 Kone MonoSpace elevator

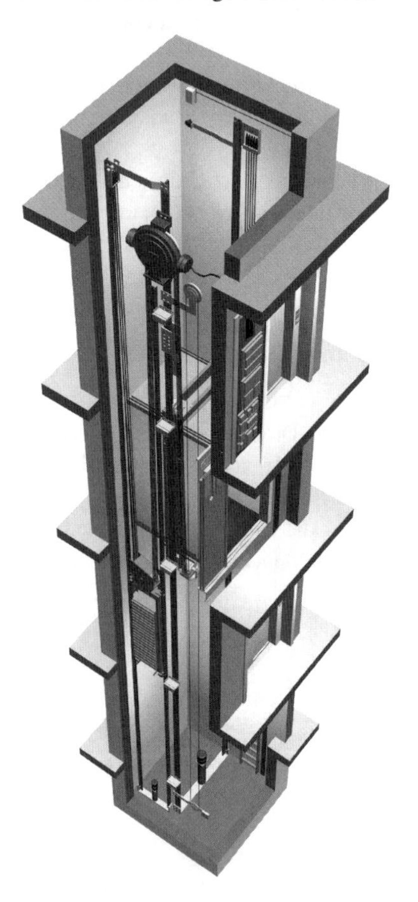

5.1.3 Bass guitar designed by Jean-Remi Corti. The material content is reduced to the bare essentials and everything not strictly functional has been simply written off. The body hosts the electronic components: natural amplification is kept to a minimum and inside there is only the pick-up (a sort of microphone that draws in and transmits the vibration of the cords to the amplifier) and a mini-amplifier that allows direct connection with a pair of headphones.

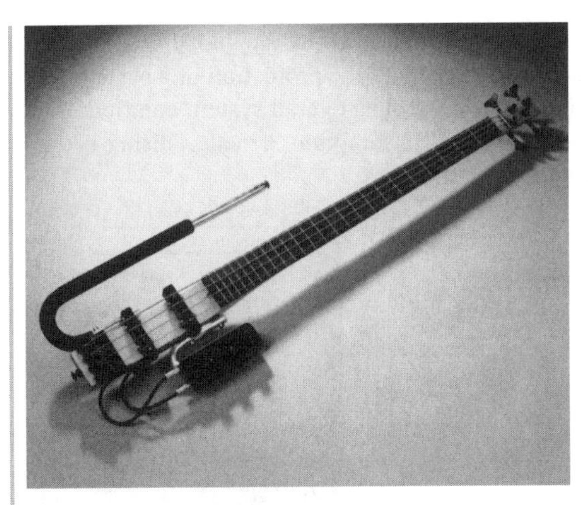

Example 5.1.3 Bass guitar by Jean-Remi Corti

5.2.1 Card payments substitute the paper and metals used for notes and coins, dematerialising the payment operation.

5.2.2 Various phone companies offer centralised answering machine services. Thus, no user actually has to own an additional appliance for the phone.

5.2.3 Sharp manufactures a series of electronic gadgets called E-dictionaries. Digitalisation reduces the required materials (paper and ink); it should also be underlined, that updating such a product does not require reprinting.

Example 5.2.3 Electronic dictionary by Sharp

5.2.4 The iPod is a digital audio player marketed by Apple Computer since October 2001. Its marketing is complemented by the promotion of a system and a website to download audio files on demand. The overall system contributes to minimising material content within music consumption, dematerialising listening devices (tapes and CDs) and their distribution.

Example 5.2.4 iPod MP3 player, Apple

5.3 Novembal bottle cap for non-gaseous liquids has a closing method (an entrance capsule) that reduces material requirements. It has a triad thread instead of a traditional one and its external measurements are 26.7 mm instead of the standard 28 mm. The cap weighs 1.5 g and the capsule 3 g, while the standard weights are respectively 3 g and 6 g. Substituting the traditional cap would reduce 50% of the required HDPE (usually over-sized for caps of non-gaseous drinks).

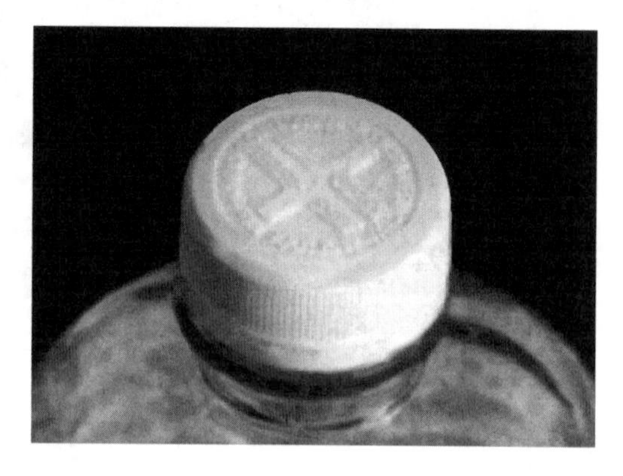

Example 5.3 Lighter bottle cap by Novembal

5.4 Valcucine has developed multi-layered doors comprising a structural chassis and pane. The frame is made of an aluminium wire and has minimal cell walls, without compromising its resistance. The pane, made of wood or laminated HPL, measures just 2 mm in thickness and is constructed using the latest pressing and deformation technologies studied by the car manufacturing sector.

The layers are lightly joined, a thin pane (laminated, in glass or marble) on a structural profile made of aluminium. The structure allows a drastic (85%) reduction in materials employed.

It is also possible to save up to 70% of the raw materials used for the frame (made of glass and aluminium) using an assembly system that removes double sides, combining different modules and integrating the feet inside the frame (eliminating the doorsill).

Example 5.4 Multi-layered, lightweight door by Valcucine

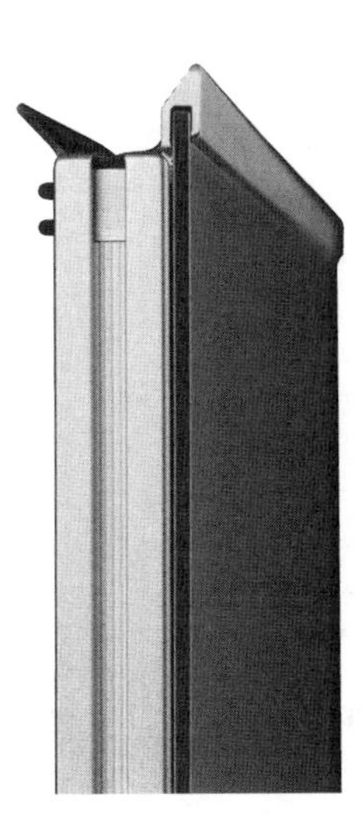

5.2.2 Minimising Scraps and Discards

A correct systematic vision of product material consumption refers not only to the material content in the final product, but also to the materials necessary for manufacturing, including the scraps and discards of the production stage and resources consumed by various operations during the transformation of its components and materials into components.

Guidelines for minimisation of scraps and discards:

- Select processes that reduce scraps and discarded materials during production (Example 5.5)
- Engage simulation systems to optimise transformation processes (Example 5.6)

Examples

5.5 Compwood Machine Ltd. has conceived a pre-compression technology (Compwood™) for solid wood before bending processes. Timber is soaked, heated to 100°C and plasticised. After heating the wood is pressed down to 80% of its original length.

When the pressure is removed the wood will regain most of its length, but will lose about 5% permanently. All its axial fibres are bent like in an accordion. The wood can then be bent in any direction and fixed during the drying process.

Traditional vapour-based bending allows a smaller range, but this operation can be carried out while cold or hot. The discards and operational stages are minimal, the equipment less complex and lighter than in vapour-based processes.

Example 5.5 Wood-bending technology, Compwood

5.6 Simulation of inkjet printer injection with computerised analytical systems can further the optimisation of various parameters, helping to minimise the energy consumption, never mind the discards.

5.2.3 *Minimising Packaging*

Proposing the reduction of materials in packaging does not mean that we should underestimate what they do to keep the product intact during transportation and storage. Thus, the packaging can have an environmental advantage because they increase the average life-span of a product, preventing it from being damaged.

But otherwise we have underlined[6] that packaging should be considered as a product in its own right, with a proper life cycle. Therefore, many of the strategies proposed also deal with the packaging. Here, three particular indications are given.

[6] *Cf.* Chap. 4.

Guidelines for packaging minimisation:

- Avoid packaging (Examples 5.7)
- Apply materials only where absolutely necessary (Examples 5.8)
- Design the package to be part (or to become a part) of the product (Examples 5.9)

Examples

5.7.1 Celaflor has substituted its former blister packaging of gardening products for a new container that has the refills fitted into an internal cavity, thus effectively avoiding multiple packaging.

Example 5.7.1 Packaging for gardening products, Celaflor

5.7.2 Lancôme has partly removed its cream packaging, reducing the package weight by 40%.

Example 5.7.2 Packaging for cosmetics, before and after redesigning, Lancome

5.7.3 Tertiary packaging and pallet stabilisation methods can be designed in a way that avoids using the internal layers that are normally employed for granulating.

5.7.4 Lush shampoo is being sold in shops per weight, drastically reducing the need for packaging (secondary and tertiary). When purchased, the shampoo is transported in a recycled paper bag.

Example 5.7.4 Unpacked Lush shampoo

5.8.1 In recent years the plastic thickness in bottles has been notably reduced thanks to advanced ribbing, which allows the necessary rigidness to be preserved.

5.8.2 The packaging for a bicycle has been maximally reduced; only the bare minimum for protection and transportation remains and material consumption appears only where absolutely needed.

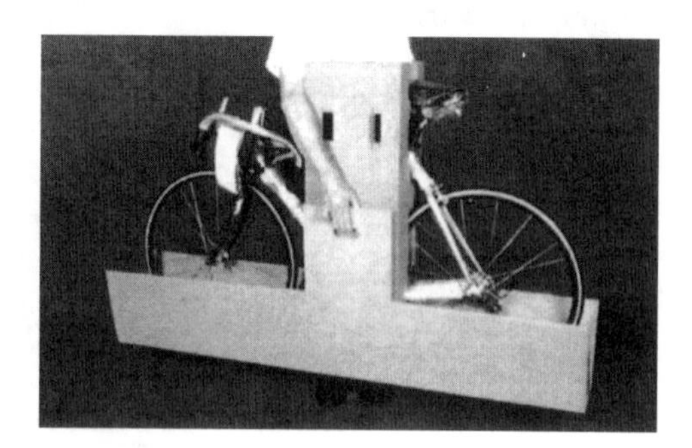

Example 5.8.2 Mini-packaging for a bicycle

5.8.3 Domus Academy has developed TV set packaging where internally expanded polystyrene parts are removed to reduce the overall dimensions. Only the corner bits have remained and the package is fastened with an external cardboard layer (that would have been part of the product wrapping anyway).

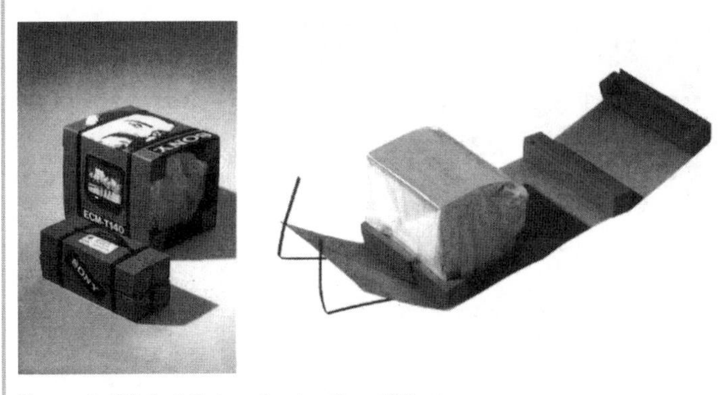

Example 5.8.3 Mini-packaging for a TV set

5.9.1 Polymer producer Dupont has adapted for elastomer (neoprene) packaging a compatible membrane (still made of neoprene) that does not have to be removed before usage.

5.9.2 Eckes chocolate wrappings are made of edible materials (wafer). They transform from packaging into product and there is no need to throw them away.

Example 5.9.2 Edible packaging for chocolates, Eckes

5.9.3 Tupa bamboo beds, designed by Paolo Balderacchi and Massimo Gregoricchio, are sold in a canvas bag that can be turned into the foothold of the bed itself: the bag is cut into four pieces, filled with fine gravel, rice, corn or stones, and then attached to the bed.

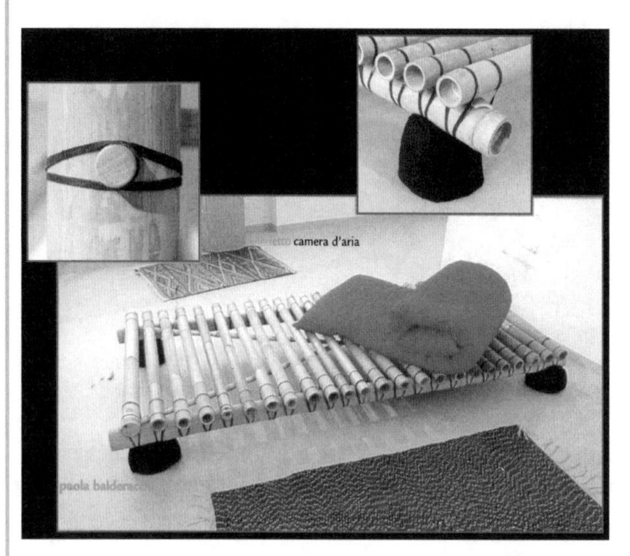

Example 5.9.3 Bed footholds for Tupa, Centro Bioedile

5.2.4 Minimising Materials Consumption During Usage

This part does not deal with general performance of the products, but only with the environmental requirements of reducing material consumption during usage, thus referring to all possible developments towards more efficient material consumption.

While it is important to adapt more efficient systems that are already available, then it is also necessary before adapting the products to understand accurately the system operation requirements, performance and consumption modalities. This can be acquired via intelligent systems that optimise consumption automatically or by products that require extra attention from the consumer.

New technologies give the foremost access to digital support and the programmable setup is one of the most important methods for the dematerialisation of consumption.

The design indications are divided into two categories:

- Select more consumption-efficient systems
- Engage systems with dynamic material consumption

Guidelines for minimising materials consumption during usage:

- Design for the efficient consumption of operational materials (Examples 5.10)
- Design for the more efficient supply of raw materials (Example 5.11)
- Design for the more efficient use of maintenance materials
- Design systems for the consumption of passive materials (Example 5.12)
- Design for the cascading of recycling systems (Examples 5.13)
- Facilitate reducing materials consumption for the user (Example 5.14)
- Set the product's default state at minimum materials consumption (Example 5.15)

Examples

5.10.1 The Dolomite toilet bowl has been designed to use an innovative hydrodynamic solution that focuses on efficient water consumption; 3.5 l instead of common consumption from 6+ litres.

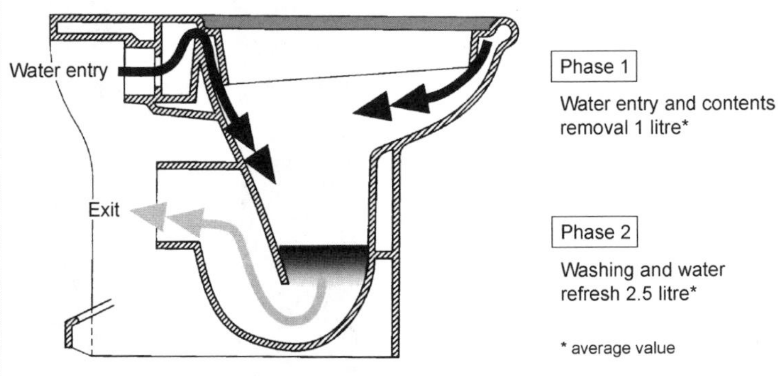

Example 5.10.1 Optimised water consumption of the Dolomite toilet set

5.10.2 The E-cloth System is a housecleaning cloth that, thanks to special microfibres, cleans the surfaces (of glass, stainless steel, chrome steel *etc.*) without chemical detergents.

Example 5.10.2 House-cleaning E-cloth system, by Enviro Systems

It is made of a single-threaded fabric of polyester (70%) and polyamide (30%), which has a great electrostatic charge capable of drawing in dust and all kinds of dirt; soaked or dry, this cloth is capable of cleaning every surface. After usage it has to be rinsed with lukewarm water and Marseilles soap.

5.11 The Sundance farm in California has engaged an underground irrigation system (20–25 cm) that issues water directly to the roots of plants. The consumption reduction results are between 65–90%. Besides that, the smaller water spills reduce the growth of weeds and consequently the use of herbicides. Also, the fertiliser is carried to the plants via these tubes at reduced rates.

5.12 Domestic irrigation often uses water from waterworks, wasting a resource with considerably higher quality than needed. Orso Raincolumn, designed by Marc De Jonghe (Orso Design Studio), uses rainwater from a collection tank.

Example 5.12 Rainwater tank, Studio Orso Design

5.13.1 Huib Van Glabeek has designed a hybrid system of washbasin and toilet bowl. Water used in the washbasin is collected and later used to flush the toilet.

Example 5.13.1 Integrated washbasin and flush toilet, Huib van Glabeek

5.13.2 Electrolux United Design has designed a hybrid system of washing machine and toilet bowl. This system has a cascade use of water. Water used by the washing machine is collected into an internal tank and later used to flush the toilet.

Example 5.13.2 Integrated washing machine and flush toilet, Electrolux

5.14 Ecostop is an internal cartridge made of ceramic discs and installed into several Grohe water taps. The device divides the opening positions into two. Opening the tap will start at the *economic* position (0–5 l/min), and can be used for smaller necessities. For higher demand, the tap can be opened to the *comfort* position (5–13 l/min) to employ the maximum supply. This feature helps to reduce wastage of water.

Example 5.14 Dual control water tap, Grohe

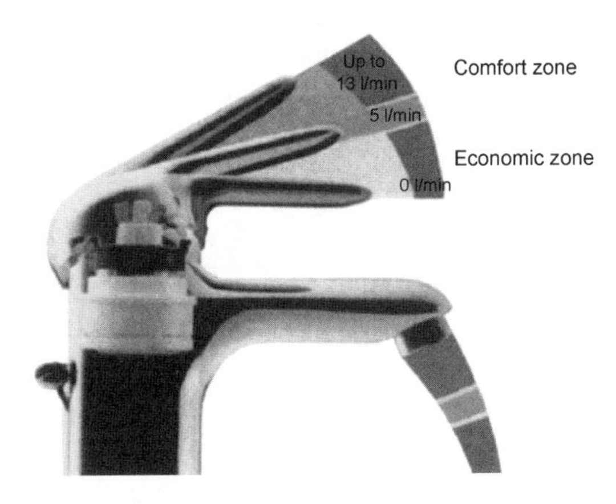

5.15 Photocopiers should have their default state reprogrammed to double-sided printing.

Guidelines for engaging systems with dynamic material consumption:

- Engage digital support systems with dynamic configuration
- Design for dynamic material consumption according to differentiated operational stages (Example 5.16)
- Engage sensors to adjust materials consumption according to differentiated operational stages (Example 5.17)
- Reduce materials consumption in the product's default state

Examples

5.16 Cesame is marketing a toilet bowl with a dynamic flush. This allows water consumption to be reduced, choosing the necessary option before flushing.

Example 5.16 Dual control flush toilet, Duetto Cesame

5.17 Izzi dishwasher, designed by Roberto Pezzetta, makes use of *fuzzy* logic technology to inform the appliance via a set of sensors how to optimise the operational stage automatically and autonomously. Based on the load and its dirtiness, the machine selects automatically the best available program, the temperature, the amount of water and the length of washing time, reducing resource consumption to a minimum.

Example 5.17 Fuzzy logic dishwasher, Rex Izzi

5.2.5 Minimising Materials Consumption During the Product Development Phase

Also, product development designing processes call for the minimisation of materials consumption. Information and telecommunication technologies allow information to be designed and exchanged with great efficiency, without wasting material resources such as paper and ink. These opportunities have a clear environmental advantage, but are often also more comfortable for the users.

Guidelines for minimising materials consumption during the product development phase:

- Minimise the consumption of stationery goods and their packages (Examples 5.18)
- Engage digital tools in designing, modelling and prototype creation (Example 5.19)
- Engage digital tools for documentation, communication and presentation (Examples 5.20)

Examples

5.18.1 Make two-sided photocopies.

5.18.2 Print texts with single line spacing.

5.18.3 Circulate articles and memos in rotation to reduce paper consumption.

5.18.4 Recycle the paper used for notes and memos.

5.18.5 Use adhesive labels for fax instead of whole paper sheets.

5.18.6 Buy envelopes without cellophane or other extra packaging.

5.18.7 Recycle printer toners and cartridges.

5.19 Engage digital modelling software to visualise and render digitally the aesthetic characteristics of the product.

5.20.1 Use e-mail.

5.20.2 Use the internet and LAN for document shipments.

5.2.6 Minimising Energy Consumption

This paragraph illustrates with examples the guidelines for minimising energy consumption for and by the product, including the pre-production, production and transportation stages. Particular focus will be on the following guidelines:

- Minimising energy consumption during pre-production and production
- Minimising energy consumption during transportation and storage
- Selecting systems with an energy-efficient operation stage
- Engaging the dynamic consumption of energy
- Minimising energy consumption during product development

5.2.6.1 Minimising Energy Consumption During Pre-production and Production

Energy reduction during these stages refers to interventions that reduce and optimise the consumption during all operations that relate to the production of the product and its component materials, starting from employing materials with lower energy intensity, optimising the parameters for productive processes, to the more efficient storage and transportation structures for materials and sub-assembly, reduction of discards from miscalculated storage volumes and deterioration of old stock, right up to efficient heating, air conditioning and lighting structures for industrial buildings.
Here, the designer can intervene with a certain incisiveness in the selection of materials and processes with low energy consumption.

Guidelines for minimising energy consumption during pre-production and production:

- Select materials with low energy intensity (Example 5.21)
- Select processing technologies with the lowest energy consumption possible
- Engage efficient machinery
- Use the heat emitted in processes for pre-heating other determined process flows
- Engage pump and motor speed regulators with dynamic configuration
- Equip the machinery with intelligent power-off utilities
- Optimise the overall dimensions of engines
- Facilitate engine maintenance
- Define accurately the tolerance parameters
- Optimise the volumes of required real estate
- Optimise stocktaking systems
- Optimise transportation systems and scale down the weight and dimensions of all transportable materials and semi-products
- Engage in efficient general heating, illumination and ventilation in buildings

Example

5.21 Aluminium production consumes a great deal of energy, especially when compared with other materials or recycled aluminium; the latter allows a reduction of approximately 90% (aluminium is necessary for highly mobile or frequently transported products.

5.2.6.2 Minimising Energy Consumption During Transportation and Storage

Concerning transportation it is clear that means of transport with lower impact have to be selected[7]. For this, the energy consumption of transportation has to be minimised. Besides, the designer can choose solutions that maximise the vehicle and warehouse capacities and thus minimise the consumption rate per transported unit. Reconfiguration of packages and goods can increase the rate of space per transported goods.

On the other hand, transportation does not concern only the products, but also the materials and energy required for the functioning of transportation and related processes. In any case it is preferable to choose local energy sources, which also has a considerable impact on the socio-political profile of sustainability, because it favours the economical and environmental sustainability of local communities[8].

Guidelines for minimising energy consumption during transportation and storing:

- Design compact products with high storage density (Examples 5.22)
- Design concentrated products (Example 5.23)
- Equip products with on-site assembly (Examples 5.24)
- Scale down the product weight (Example 5.25)
- Scale down the packaging weight
- Decentralise activities to reduce transportation volumes (Example 5.26)
- Select local material and energy sources (Examples 5.27)

Examples

5.22.1 B&R Meyer Gmbh has developed a shampoo packaging, where the cap of the bottle fits into the base of another bottle, and has rectangular shape. These characteristics combined reduce about 40% of transportation volumes.

[7] *E.g.* choosing railroad transport instead of trucks.
[8] This topic has been already discussed in Part I of this book.

5.22.2 Eco chair, designed by Peter Karpf, is made of a single piece of beech wood, bent three times for its angles and has a cleavage cut that forms the thin legs. The chairs are highly stackable and grant optimum conditions for storing and transportation.

Example 5.22 Eco chair by Futureproof

5.23 Procter & Gamble, like many other producers, market concentrated washing powders and conditioners. Results are very positive, both economically and ecologically, softener Lenor Ultra for example uses a formula with 6 times higher concentration than traditional products.

Example 5.23 Packaging for conditioners before and after re-designing, Procter & Gamble

5.24.1 Ikea has achieved a considerable success selling in innovative ways the furniture for homes. Some parts of their strategies have a formidable environmental effects. For example they offer home-assembled furniture that have been designed to have minimum transportation volume, this has lowered the transportation costs and won them even more customers, nonetheless reduced the environmental impact.

Example 5.24.1 On-site assembled chairs, Ikea

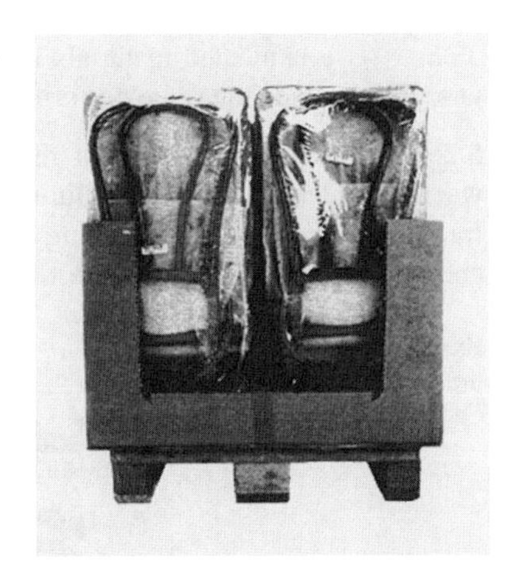

5.24.2 Gaetano Pesce designed back in 1969 the product line of UP chairs. The chairs were vacuum-packed and acquired their form when unpacked. This reduced drastically the transportation volumes.

Example 5.24.2 UP chairs by Gaetano Pesce

5.25 Use laminated materials instead of glass and foils used in bottles and jars. Will provide the same performance at considerably lower weight.

5.26 New electronic meters allow a remote control for monitoring the networks (to identify possible failures), reading the meters and manage the contracts (new connections, finishing contracts, replacements *etc.*). In this way the energy supplier technicians don't have to pay constantly home visits.

Example 5.26 Electronic controller of electricity supply, Enel

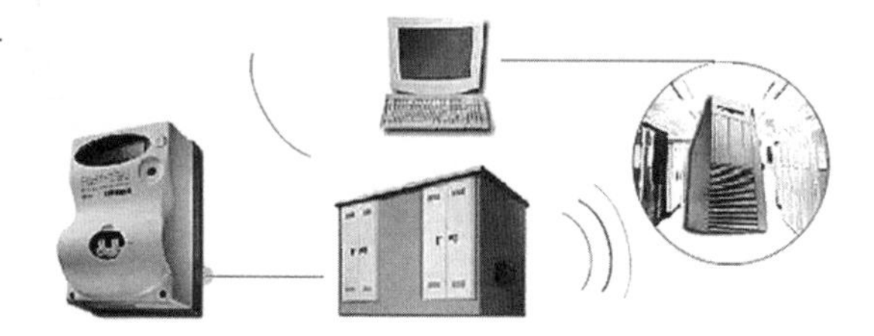

5.27.1 Solar cooker is a clean, safe and cheap cooking system, easy to build and use. Made of a parabolic reflective panel and of a plastic bag, it can be used with any pot for cooking. In many sunlight rich countries it can save for families up to half the combustibles used for cooking, reducing especially the consumption of firewood and avoid deforestation. It can also used for water purification, thus reducing the risk of many diseases that are tragically typical among children of many poor countries.

Example 5.27.1 Solar-powered cooker

5.27.2 In cooperation with Industrial Design Department of Paraiba federal University in Brazil, was developed a pedal-powered washing machine for people in north-eastern regions of Brazil. In these places the laundry is normally washed manually by women, carried to appropriate buildings. This project was created to react to the local demand and its first objective was improving the working conditions, that cause serious back problems and pain. Work was carried through in close cooperation with local businessmen, renowned for their capacities to create products from scraps and discards (not only decorative, but also primary products, that later are sold to poorer part of the population). Materials are normally easily repairable on-site. The first series of prototypes were tested by some of the women and showed considerable improvement.

Example 5.27.2 Pedal-powered washing machine

5.2.6.3 Selecting Systems with Energy-efficient Operation Stage

This part does not deal with the general performance of the products, but only with the environmental requirements of reducing energy consumption during usage, thus referring to all possible developments towards more efficient energy consumption.

Similarly, what has been said about material consumption, it is also important here to adapt more efficient systems that are already available, and on the other hand it is necessary before adapting the products to accurately understand the system operation requirements, performance and consumption modalities (see Sect. 5.2.4).

Just as important are here strategies that indirectly reduce consumption. For example, designing lighter products, which, besides reducing materials consumption, lower the energy consumption in the case of transportation.

Another issue here concerns the strategies and design proposals aimed at collective usage; energy reduction in these cases can be very significant (Box 5.1).

Box 5.1 Collectively Used Products

Let us compare a collective use that responds to the social demand for results with the results offered by products, owned and used individually, and establish an hypothesis on the parity of the performances offered (parity with the demand for results by a certain number of customers).

Simultaneously operating with more customers and with a greater amount of performances demanded, the collectively used products lead to an optimisation of consumed resources and reduction of waste. This happens for two fundamental reasons: the economies of scale and greater professionalism of the operators engaged in providing the service.

Besides, the collectively used products allow easier adaptation to more advanced technological solutions in terms of consumption efficiency during use. In fact, these superior systems often have higher costs and are not economically affordable on an individual level.

Furthermore, the overall number of products required simultaneously for satisfying the demand of a community usually diminishes: if one product gives the desired results for more than one person, then fewer products are required in the given place and time.

In the end, we can reckon that even if the products in collective use might be looked after less carefully (sometimes), then at the same time the service would be supplied by professionals with greater care and efficiency in maintenance and repairs.

Here, the design guidelines are subdivided into two parts:

- Select systems with an energy-efficient operation stage
- Engage dynamic consumption of energy

Guidelines for selecting systems with an energy-efficient operation stage:

- Design attractive products for collective use (Example 5.28)
- Design for energy-efficient operational stages (Example 5.29)

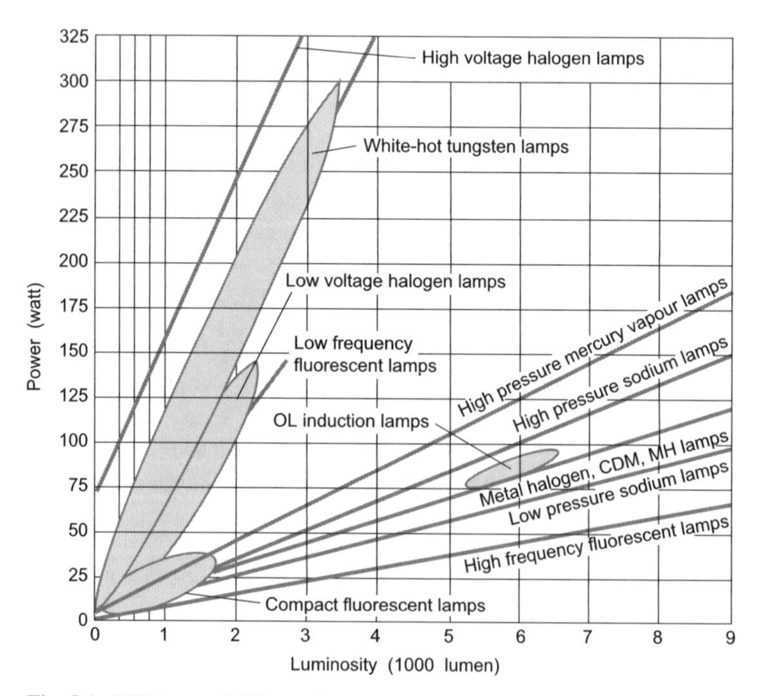

Fig. 5.1 Efficiency of different lamp types

- Design for energy-efficient maintenance
- Design systems for the consumption of passive energy sources (Examples 5.30)
- Engage highly efficient energy conversion systems (Examples 5.31; Fig. 5.1)
- Design/engage highly efficient engines (Example 5.32)
- Design/engage highly efficient energy power transmission (Example 5.33)
- Use highly caulked materials and technical components (Examples 5.34)
- Design for localised energy supply
- Scale down the weight of transportable goods
- Design energy recovery systems (Example 5.35)
- Design energy-saving systems (Example 5.36)

Examples

5.28 The city of Brescia, Italy is well furnished with an intelligent street-light system and preferential street-lanes for public transport to speed up public transport circulation. In this way, public transport has been revitalised. The buses have adjustable height to facilitate the entrance for wheelchair and push-chair users.

5.29 The Daewoo Air Power Z washing machine, designed by a group of 88 researchers, has new washing technology based on cold water. This machine washes all types of fabric with a mixture of cold water, detergent and air bubbles. During one wash up to 5 million bubbles are created, which help the washing powder to remove dirt from fibres. Delicate and coloured laundering is facilitated by the cold water, and the results are extremely good.

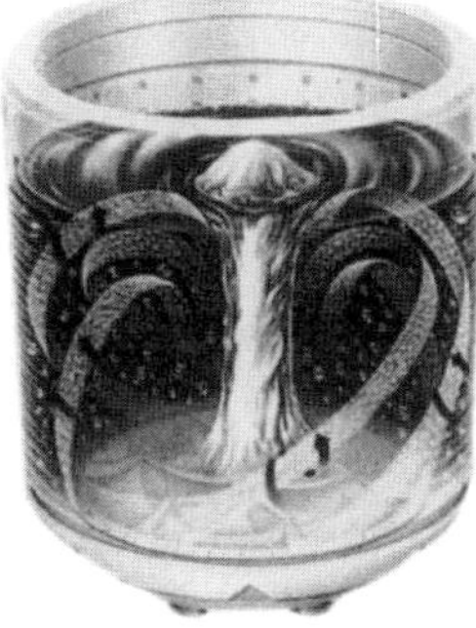

Example 5.29 Air bubble washing machine, Daewoo

In this way the laundry can be done without heating the water, which normally uses most of the required energy and deteriorates the fibre, resulting in matted wool. The Air Power Z works on only 150–200 kWh, instead of the 1,500 kWh required by common washing machines. They are equipped with a nocturnal program that works so silently that it can be used during the night.

5.30.1 The Refrigerator Fria, designed by Ursula Tischner, is meant to be installed permanently. Conceived as an architectural element, it is built into the wall, preferably into the external wall, to use the winter cold. It was calculated that in typical German house it could work for about 3–5 months a year without consuming any energy. The Fria is a multi-compartmental refrigerator equipped with up-to-date cooling and isolation systems. The standard model from 1994 uses only 0.4 kWh within 24 h; meanwhile, common refrigerators use about 0.85 kWh.

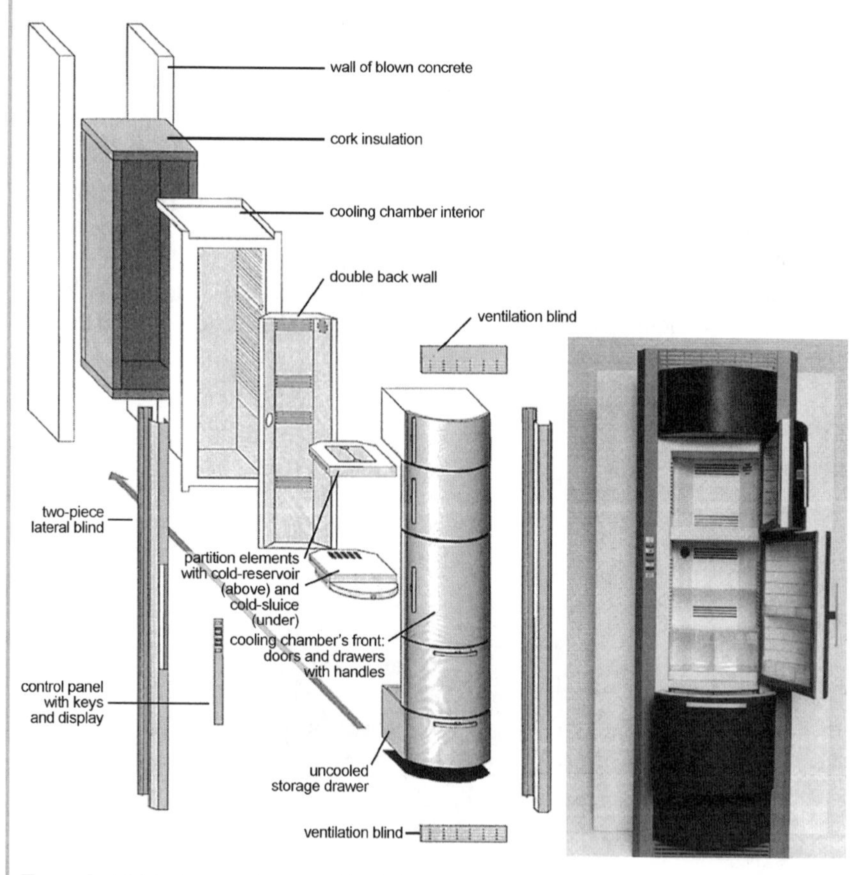

Example 5.30.1 Fria refrigerator, Ursula Tischner

5.30.2 Passive house of Darmstadt, Germany is a interesting pilot project. *Passive* refers to the constant use of solar energy, the house has almost no active central heating at less than 15kWh/m^2 per annum. Result on this scale is obtained with efficient isolation systems, both of the windows and walls.

Example 5.30.2 Passive house in Darmstadt

5.31.1 Mix desk lamp, designed by Meda & Rizzato, is equipped with LED *chip on board*. This, combined with the brightness enhancing lens and coloured filters to regulate intensity of colours, allows the LED light to be used in desk lamp, with a electricity consumption at mere 5W.

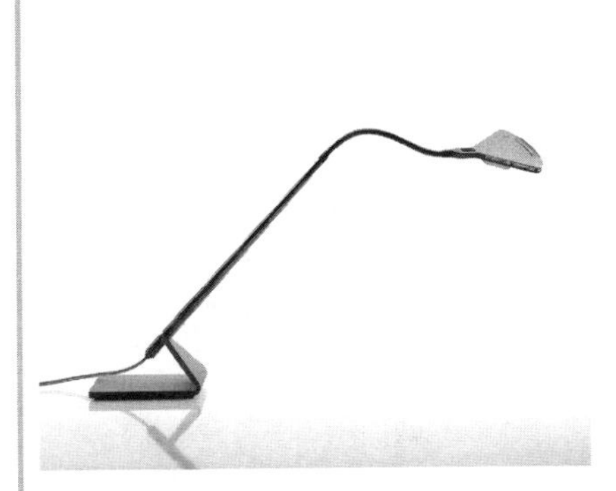

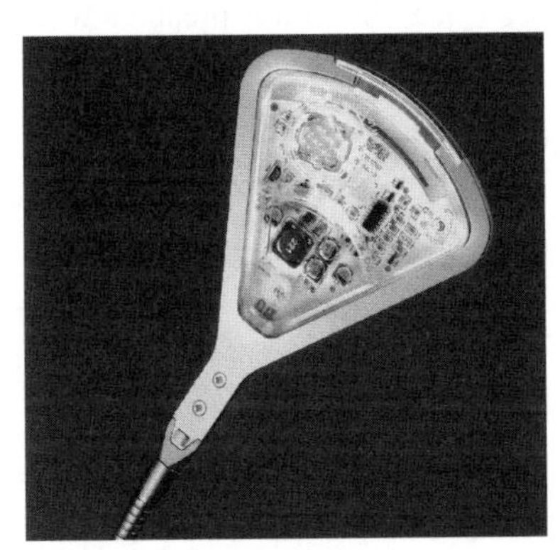

Example 5.31.1 LED Mix desk lamp, Meda & Rizzato, Luceplan

5.31.2 Solar energy research group in Nuremberg, Germany, studied the possible application of photovoltaics for average German households. According to them, a 8m^2 solar panel (together with passive water heating) should be enough to supply the appliances with direct current. Meanwhile the appliances that require alternating current also require about 30m^2 solar panel. This solution could be employed also in developing countries, where the system supplied by alternated current has not been fully introduced.

5.32 Toyota Prius is the first car in the Hybrid Synergy Drive series, a system that combines an electric engine (68 CV, zero emissions, autorecharging batteries) with a 1500 cm^3 petrol engine (78CV). Two engine work in synergy and offer an optimal performance with low consumption (4.2 l/100 km out of town).

Example 5.32 Toyota Prius

5.33 3M has been marketing for years an innovative solution for light. Scotch light conveys light via absolute internal reflection (similar principle is employed in optical cables). Scotch light is practically a flexible film with a prismatic structure, on one side of the film there is another similar one. The source of the light can be a small lamp or sunlight, later the light is released in zones where the film is prepared with bigger granules.

Example 5.33 Scotch light, tubular street light by 3M

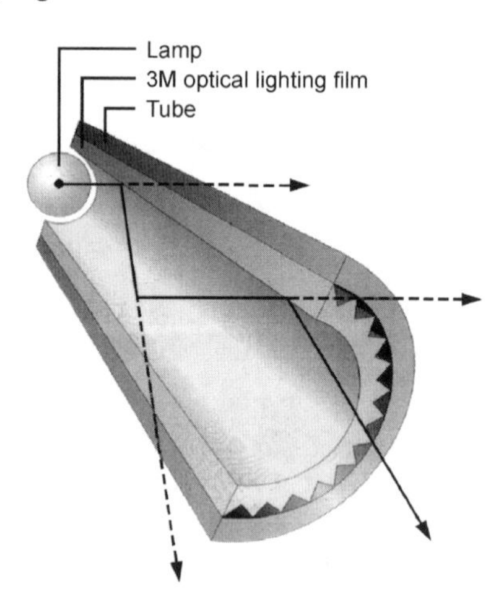

5.34.1 One of the most efficient methods for caulking (even if it costs more) is obtained with two layers of stainless steel, distanced at few mm and welded at sides. The internal space would be air-free and covered film that reflects infrared rays.

5.34.2 Superwindow is made of glass that is covered with high-tech transparent layer, which reflects infrared rays. They're available in many different characteristics according to need. This system can be employed for example for windows, where infrared rays are blocked, but the luminosity passes nevertheless. In Rocky Mountain Institute in Colorado, this technology has been combined with isolation, where space between glass layers has been filled with heavy gases that enhance this system even further.

5.35 i-Magic Fortius is a bicycle, that recovers muscular energy released during the pedalling and transfers it to network. It saves the energy that would have been wasted otherwise.

Example 5.35 i-Magic-Fortius, bicycle that recovers muscular energy

5.36 Many cars are equipped with indicators showing the petrol consumption during driving; these indicators motivate drivers to pace at lower petrol consumption.

Guidelines for engaging dynamic consumption of energy:

- Engage digital dynamic support systems
- Design dynamic energy consumption systems for differentiated operational stages
- Engage sensors to adjust consumption during differentiated operational stages (Examples 5.37)
- Equip machinery with intelligent power-off utilities (Examples 5.38)
- Program the product's default state at minimum energy consumption

Examples

5.37.1 Luminous traffic signs, designed by the *nlplk* design studio, Netherlands, are equipped with fluorescent lights and sensors; the latter increases the energy supply at lower temperatures (when the efficiency of the fluorescent lights decreases). Energy saving, about 40%.

Example 5.37.1 Luminous traffic signs by nlplk

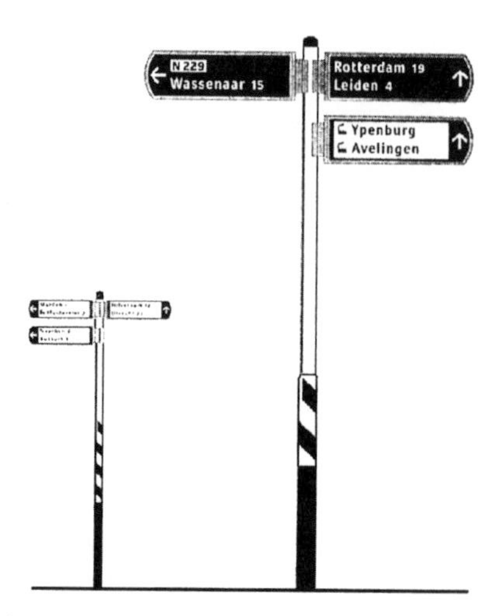

5.37.2 Luxmate daylight provides artificial light according to the presence of natural light. Depending on weather conditions, every room is lighted more or less by natural light. Luxmate daylight measures the presence of natural light with sensors. The lights near the window will dim themselves down first (down to 1% and then eventually turn themselves off); consequently, other lights react similarly. The other way round, if the outside light supply diminishes, artificial lighting returns to higher levels (according to the system installed either automatically or from the switch).

If there is a need for extra light in addition to the programmed amount, it can be provided with a corresponding switch.

Luxmate daylight grants a possible saving of between 60 and 75%.

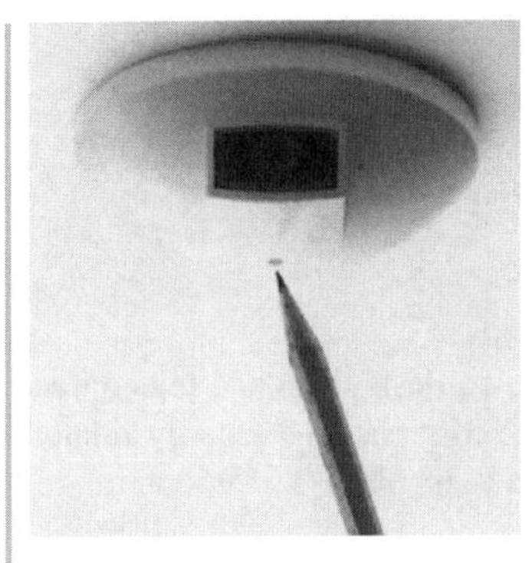

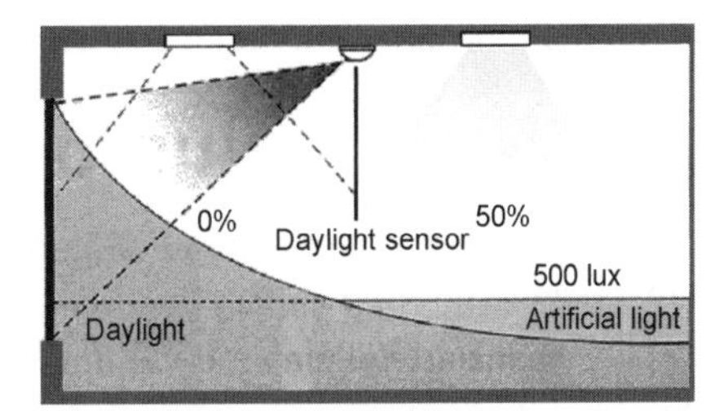

Example 5.37.2 Luxmate daylight, light with external sensors, Zumtobel

5.38.1 Philips radio SBC SC 390 is equipped with a sensor that switches the radio on when someone enters the room. The sensor is made of a photocell that activates the radio, when someone passes within its range. This product allows electricity consumption to be minimised via an automatic on–off switch, avoiding the radio playing idly, without actual demand (or when there is nobody to listen to it).

Example 5.38.1 Philips Radio with movement sensors

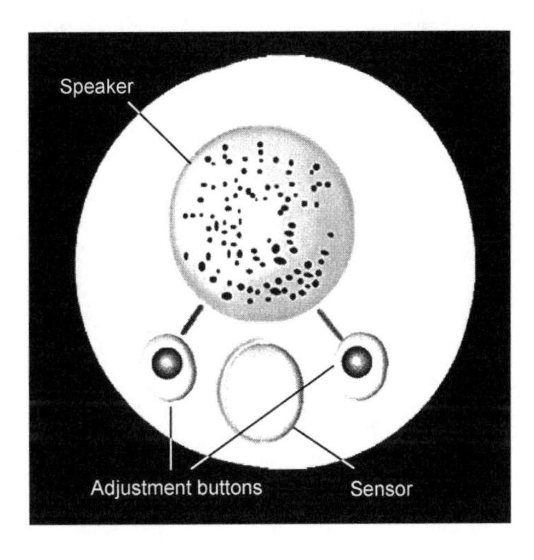

5.38.2 Many computers have an automatic break contact function that switches off the system when not in use.

5.2.6.4 Minimising Energy Consumption During Product Development

Also, product development design processes call for minimisation of energy consumption. Information and telecommunication technologies allow design and exchange of information with great efficiency, without demanding great mobility of goods and people. These opportunities have a clear environmental advantage, but are often also more comfortable for the users.

Guidelines for minimising energy consumption during product development:

- Engage efficient workplace heating, illumination and ventilation
- Engage digital tools for communicating with remote working sites

Examples

5.39 Telecommuting (including long-distance co-designing) and teleconferences are potentially a phenomenal source of economic and energy-related savings: eliminating the need to move from place to place and collateral resource consumption.

Chapter 6
Selecting Low Impact Resources and Processes

6.1 Introduction

Design choices should be aimed at resources (materials and energy sources) with lower impact, while offering equal terms of service or *functional unit* with the life cycle of a product.

The designer plays an important part in choosing and using materials, even if he/she is generally not concerned where they come from or where they end up after the life cycle ends. The same is true with regard to the selection of energy sources that are necessary for the product's operational stage. A bit less incisive is his/her role during the production and distribution stage, but an informed designer can also offer low impact alternatives here as well.

Is important to remember that for an accurate and efficient approach to reduction of environmental impact, it is necessary to redesign the entire product system as much as possible. Every calculation has to refer to the whole life cycle and all its processes.

This means that we have to choose between different transformation technologies (some of them might create toxic and harmful emissions, some, equally efficient, not); choose distribution structures that cause less harm to nature[1]; design products that use fewer resources (energy and material consumption) with less impact; in the end we have to orientate the selection of materials and additives in a way that minimises the risk of harmful emissions during end-of-life treatments. But the calculations for pre-production must also take into consideration the working environment and its risks.

Finally, from the environmental sustainability perspective, which aims to save resources for future generations as well, their renewability,[2] or at least their non-exhaustibility, gains greater importance.

---

[1] Generally, rail-transport should be preferred to other land transport or air transportation.
[2] *Cf.* Chap. 12.

The selection of low impact resources is, let us say, a qualitative saving for nature that can be achieved in two conceptually different ways:

- Selection of non-toxic and harmless resources
- Selection of renewable and bio-compatible resources

6.2 Selection of Non-toxic and Harmless Resources

In the interests of clear understanding and an efficient supporting structure for designers, the guidelines are divided according to the nature of the resource:

- Selection of non-toxic and harmless materials
- Selection of non-toxic and harmless energy resources

6.2.1 Select Non-toxic and Harmless Materials

Products consist of materials, therefore the materials, as they form the components of a product determine the various environmental impacts and their effects on the health of us and our ecosystem.

The production of materials (pre-production stage), starting with the extraction of raw materials, consumes energy and other raw materials, and effectuates emissions. Several methods have been developed that calculate the environmental impact of different materials according to the rate of weight per production process.

The bar chart of Fig. 6.1 shows the estimated environmental impact of some commonly used substances according to the method of the Ecoindicator[3]. The charts can show important information about the harmfulness of the production of different materials, but it is not enough. Selection may not disregard the impacts, but neither can it ignore the possible environmental advantages that might result during other than the pre-production stages.

In the end, all materials have an impact on nature, some more, some less, but to compare them the function and services that products, not only materials, offer, has to be taken into consideration.

A certain substance might have a worse impact during production and disposal than others, but it might expand the life cycle of the product in time or in efficiency. Besides, if a product had a longer lifespan, it would avoid the pre-production, production and distribution of new components for substituting prod-

[3] Values are calculated at the impact effectuated by the production of 1 kg of a given substance. The estimations method has been developed by the CML centre of Leiden University in collaboration with the Netherlands government, Philips, Oce, Nedcar and Fresco. Its criteria are described in detail in one of the examples in Chap. 12.

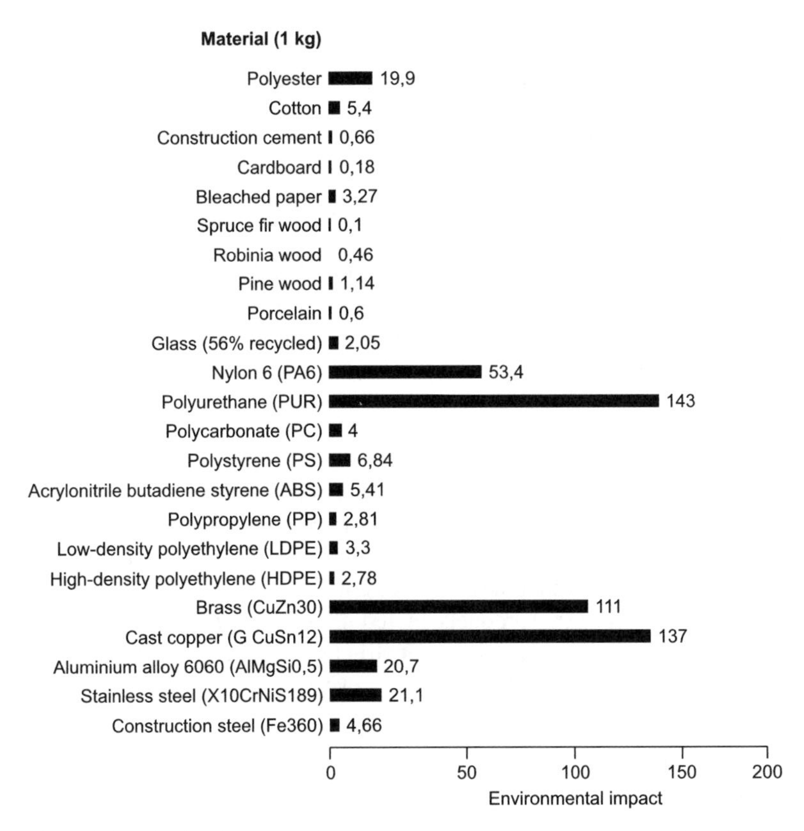

Fig. 6.1 Production impact index of some common materials (according to the method of Ecoindicator 95)

ucts[4]; or if the material allowed reduced energy consumption, then the product would have a smaller impact altogether[5].

Besides, according to different production and consumption contexts, the impact can be different. Some regions have ample reserves of some resources and scarce reserves of others, or their production and disposal customs might be harmful to nature.

In summary, impacts can occur at all stages, during the extraction of resources and production of materials as well as during their transformation, distribution and disposal. Therefore, the choices made to minimise hazardous emissions should equally take into consideration the production processes, transformation of materials, distribution structures and end-of-life treatments. In the case of products that consume different resources during their usage, it is necessary to design them so that consumed resources would have the smallest impact possible.

---

[4] *Cf.* Chap. 7.

[5] Composite materials normally have a high impact during production and are not easily recyclable. Anyhow, given the lightness and resistance they might have, they are indispensable materials for vehicles due to the reduction in energy consumption.

If toxic and dangerous substances have to be utilised, at least the precautions should be foreseen, which would minimise the risks during all stages of the life cycle.

We can still start from the truism that not using the materials will have zero impact[6]. Besides minimising the overall amount of materials, it is also important to select single materials according to their impact. And here the rankings that lay out the environmental impacts of different materials could be useful, though it would not be wise to rely on them uncritically, without the perspective of the contexts in which they are used and the life cycles of the products that they are part of.

Some materials, regardless of the usage and place of origin, have been attested as being too dangerous substances; employment of these substances is normally prohibited by law (Example 6.1).

Example

6.1 Asbestos is a natural material that has had elevated risks throughout its lifespan. Ever since World War II, its employment has grown increasingly.

Its chemical stability and rigidity make it an optimal reinforcing agent for plastics and cement. It is also easy to mine and process at low costs, and this substance has been used in more than 3,000 products, including isolation materials and heat-resistant clothes. Recently, the demolition of structures containing asbestos, for example, for heat isolation, has turned out to be a very dangerous operation.

Exposure to asbestos can indeed cause these diseases:

- Lung diseases and cancer
- Asbestosis
- Gastric cancer
- Small intestine cancer

The cause of all this can be found in the geometry of the asbestos fibre. Their structure bestows upon this substance an extremely delicate surface that allows it easily to bond with and mutate molecules. The longer the contact with human tissue (especially lung tissue), the greater the possibility of cellular deformation.

Since 1991, many countries have taken legislative measures to prohibit its usage.

Recent studies have shown that other man-made fibres with a similar structure, such as ceramic and aramid fibres, can be equally as carcinogenic.

It often occurs that the cause of toxic and harmful emissions during production, usage, incineration and discharge is not so much connected to the materials, but rather to the additives[7]. It is preferable to avoid this kind of additive or select another material altogether (Box 6.1).

---

[6] The minimisation of materials was defined in the previous chapter.

[7] *E.g.* polymers are often complemented by additives against heat, fire and ultraviolet rays, with reinforcing, filling, weighing, expanding and antioxidant additives.

Box 6.1 New Materials

The usage of new materials has been technologically and scientifically studied with great attention. One of the main objectives of research has been the development of highly resistant materials; some studies try to improve the electric and optic performance of some substances, such as the superconductors and optical fibres. General knowledge of their properties and predictability of their performance has brought us to the threshold where materials can be designed at a writing desk in order to respond adequately to the given demand for performance. It is enough to think about all ceramic, composite and intelligent materials that are capable of adapting their function according to alterations in external conditions (Table 6.1).

On the other hand, the criteria of reducing environmental impacts are the same with new materials as they were with old materials.

Table 6.1 Advantages and disadvantages of the new materials

Material type	Ceramic	New alloys	Polymers	Compounds
Positive qualities	Heatproof, highly rigid, high thermal conductivity, corrosion-proof, thermal shock-proof	Heatproof, corrosion-proof, mechanical resistance, fatigue-proof	Lightness, corrosion-proof, mechanical resistance	Corrosion-proof, lightness, mechanical and rigid resistance, friction and abrasion resistance
(Suspected) negative environmental qualities	Possibly carcinogenic fibres, not recyclable in solid form	Recycling requires high energy consumption	Toxic raw materials (PEEK), toxic solvents, thermosetting substances are hard to recycle	Hard to separate from components, hard to recycle, create solid waste
Positive environmental qualities	Long lifespan, low degradability	Long lifespan	Lightness reduces energy consumption, long lifespan, highly recyclable, degradable	Lightness reduces energy consumption
Examples	Zirconium oxide (ZrO_2), aluminium oxide (Al_2O_3), sialon, silicon nitride, silicon carbide	Aluminium alloys Al-Li, aluminoids (Ti, Ni)	Thermoplastics: PEEK, polyamide, thermosetting substances: polyesters, siloxanes, PBT, aramids	Fibres: graphite, SiC, aramids Compound matrix: polymers, alloys, graphite

Guidelines for non-toxic and harmless material selection:

- Avoid toxic or harmful materials for product components (Examples 6.2)
- Minimise the hazard of toxic and harmful materials (Example 6.3)
- Avoid materials that emit toxic or harmful substances during pre-production (Example 6.4)
- Avoid additives that emit toxic or harmful substances (Examples 6.5)
- Avoid technologies that process toxic and harmful materials (Examples 6.6)
- Avoid toxic or harmful surface treatments (Examples 6.7)
- Design products that do not consume toxic and harmful materials
- Avoid materials that emit toxic or harmful substances during usage (Example 6.8)
- Avoid materials that emit toxic or harmful substances during disposal (Example 6.9)

Examples

6.2.1 Do not use polychlorobiphenyl (PCB) and polychlorotriphenyl (PCT) for transformers.

6.2.2 Substitute lead alloys for welding with tin, copper and silver alloys.

6.3 SafeChem, producer of chlorinated solvents, has developed a new container (Safetainer) equipped with an air compressor that avoids possible leaks of dangerous emissions during transportation and during transfer to the distilling machinery.

Example 6.4 Foxfibre cotton, used for the Esprit brand

6.4 Foxfibre cotton, marketed by Natural Cotton Colours Inc., is made of biologically cultivated cotton, with no application of synthetic pesticides. It is not bleached and requires no colouring. This cotton is used for example by Levi Strauss and Esprit.

6.5.1 Do not use polybrominated biphenyl (PBBE and PBB) fire retardants.

6.5.2 Do not use cadmium and its compounds for colouring pigments in plastics, paints or for covering metal surfaces.

6.5.3 Avoid using chlorinated paper, as its treatment waters are hard to filter. Alternative substances for bleaching are industrial ozone and oxygen.

6.5.4 Avoid using aldehyde-based adhesives, as they have high emissions of free aldehyde. Alternative adhesives are being researched, *e. g.* based on polypropylene and cellulose.

6.6.1 Prefer pleating to welding.

6.6.2 Do not use chlorofluorocarbons for polymer foaming, cleaning printed circuits or degreasing metals.

6.6.3 Prefer alkaline to halogenated fluorocarbons for cleaning alloys. Fluorocarbons are dangerous to the ozone layer.

6.7.1 Do not use pentachlorophenyl with wood products.

6.7.2 Ink cartridges have potential dangers related to emissions of volatile organic compounds (VOC), which can be avoided using water-based or vegetable inks.

6.7.3 Do not use heavy metals for colouring pigments. They can be substituted with organic pigments. However, is better to colour plastics instead of painting, to avoid emissions during these processes. If painting is absolutely necessary, in order to minimise VOC emissions, it is better to use water-based varnishes instead of technologies that employ electrostatic or ultraviolet radiation-based painting.

6.7.4 Do not use oil-based paints and mordants for wood surface treatments. Alternatives are water-based paints and varnishes or UV vulcanised varnishes.

6.7.5 Do not use toluene compounds for cleaning or varnishing products.

6.7.6 Avoid galvanic treatments, because their processing waters are hard to filter. Prefer metal or chrome varnishes.

6.8 Since 1992, Bryant & May in the UK have ceased to use dichromates, sulphur and zinc in match heads. Dichromates and zinc are toxic substances; incinerating sulphur creates sulphur dioxide, the main agent in acidification.

6.9 Avoid using chlorofluorocarbons (CFC) in cooling systems and foaming agents.

6.2.2 Selecting Non-toxic and Harmless Energy Resources

Selecting and procuring energy sources are decisions made on a national political scale, which does not mean that a designer cannot choose to design products that use one energy source instead of another, corresponding to fact that some sources have different environmental impacts.

Choices will be made for all stages of the life cycle.

For example, when designing products that consume energy while in use, it is important to estimate the available energy sources according to their environmental impacts, including muscular energy[8]. Some products require minimal power or energy supply in time, meaning that activity needs the absolute minimal effort. In these cases, it could be interesting to estimate the possibility of substituting with muscular energy the current that is normally supplied by power networks or batteries. In some cases, the economic and environmental victory is surpassed by the autonomy from enforced connection with a power network or batteries. Besides, these technologies tend to be simpler, well-known and reliable; thus, they normally last longer.

We should not forget that the transportation of energy always determines energy-related or material (*e. g.* methane) losses, so this is a convenient moment to recalculate the advantages and disadvantages of using long-distance energy sources.

Guidelines for selecting non-toxic and harmless energy resources:

- Select energy resources that reduce dangerous emissions during pre-production and production (Examples 6.10)
- Select energy resources that reduce dangerous emissions during distribution (Example 6.11)
- Select energy resources that reduce dangerous emissions during usage (Examples 6.12)
- Select energy resources that reduce dangerous residues and toxic and harmful waste (Examples 6.13)

Examples

6.10.1 Back in 1958, Thorens Rivera, of the UK, was selling clockwork shavers. Winding up eight times provided 3 min of working time. A special button released the spring and activated the shaver.

[8] To be precise, muscular energy is powered by our nutrition. And producing comestibles is not impact-free. To be even more precise, this impact depends on the type of food and its production characteristics. In sum, we can claim that a 5-km race after eating biological food has a smaller impact than after eating food that has been procured with traditional cultivation, including using pesticides, eutrophication agents and surplus irrigation.

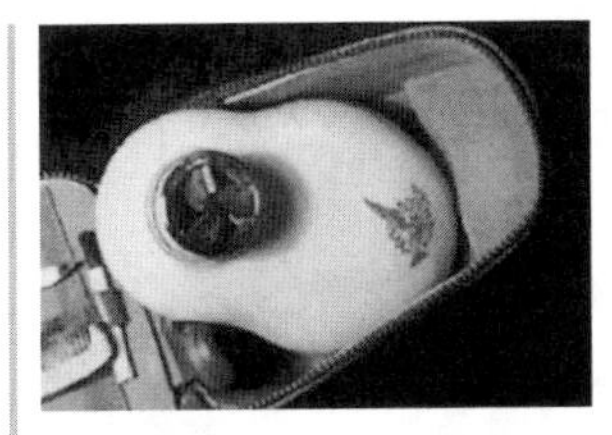

Example 6.10.1 Clockwork shavers by Thorens Riviera

6.10.2 The BayGen Freeplay radio, designed by Trevor Baylis, is a crank-up radio that is wound up with a specially designed spring and provides service on demand. Winding up 60 times (approximately 25 s) is enough for 40 min of radio service time. If the radio is switched off before the spring comes down, an electronic device saves the stored energy for later use.

The primary objective of the designer was to put at the disposal of many African populations an efficient tool to inform them better about the prevention of AIDS. In these regions, it is difficult to clamp down on the contagion of this disease, because it is also hard to organise an efficient prevention strategy. Radio could be an important instrument in this scenario if it was not for the high cost and low availability of batteries, a problem that has been overcome with this product.

Freeplay radio could also be useful in less desperate conditions in which it would give just greater freedom from the power supply.

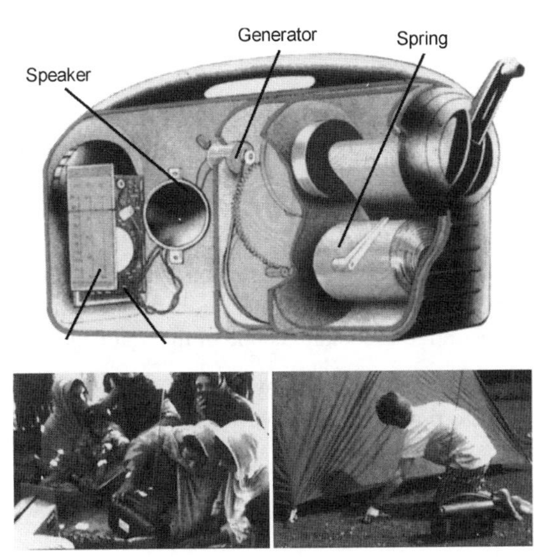

Example 6.10.2 Crank-up radio by Trevor Baylis

6.10.3 Hans Schreuder from MOY *concept & design* studio, Netherlands, has designed a clockwork toothbrush in which muscular energy is stored in a special spring.

Example 6.10.3 Clockwork toothbrush by Hans Schreuder

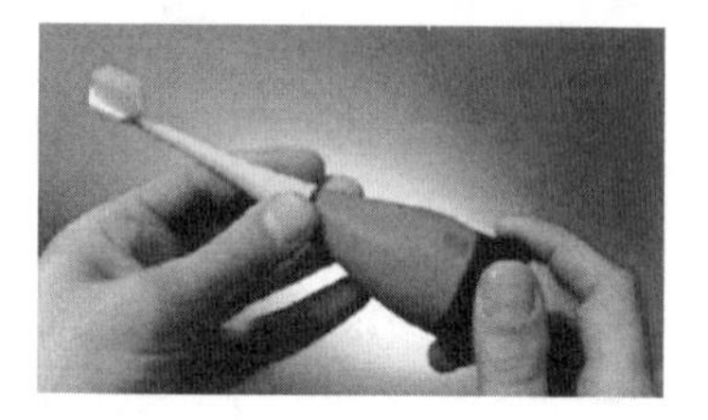

6.10.4 Since 1994, Seiko has used in its wristwatches a technology called Kinetic, quartz watches that are powered by human movement. A swinging weight is swayed by the movement of the pulse over which the watch is worn, the movement is transformed into a magnetic charge and is stored in a small condenser or a rechargeable battery that is capable of running for 5 months.

Since 1999, watches with Kinetic Auto Relay technology have been available that continue to measure time, even if not wound up for 4 years. The watch will cease to use the indicators after 70 h of inactivity, but is able to remember the correct time. When the watch is put on again the watch hands immediately start to show the correct time.

Example 6.10.4 Seiko Kinetic wristwatch

6.11 While methane is a combustible with one of the lowest environmental impacts when used (incinerated), during transportation there is a high risk of spills; in fact, methane is one of the main agents in global warming.

6.12.1 FIAT, like many other car producers, has developed car prototypes that are powered with hydrogen. Unlike engines on petrol or diesel, hydrogen emits zero toxic and harmful emissions.

Example 6.12.1 Hydrogen-powered Fiat Panda

The latest from the Turin company is the Panda Hydrogen. The fuel is stored in under-floor tanks, in which the hydrogen and oxygen react with each other, with the help of a catalyst, producing water, heat and generating energy without emissions. The hydrogen is under pressure of about 1.5 bar and together with oxygen (also under pressure) they produce energy with a maximum potency of 60 kW and with high efficiency (60% on 20% of maximum potency). Hydrogen that powers the fuel cells is kept in tanks made of compound materials, installed at the back of the car under the floor. The Panda Hydrogen has an operating range of over 200 km with urban driving. At full charge, the engine of fuel cells supplies 60 kW, allowing the car to reach a maximum velocity of more than 130 km/h, with an acceleration from 0 to 50 km/h in 5 s. Recharging time is very fast, under 5 min, and is comparable to cars that run on methane.

In 2006, FIAT started to issue small fleets of the Panda Hydrogen, as a premise, promoted and backed up by the EU, the Italian regions and the government, of greater marketing campaigns for these cars fully in 2020.

6.12.2 Tecnobus is marketing an electric minibus called Gulliver that has autonomy of 100 km and a fast recharging cycle (3 min). Like all electric vehicles, it does not emit any pollution when in use.

Example 6.12.2 Gulliver electric minibus

6.12.3 MULO (Mobilita Urbana da LavorO), is a hybrid quad cycle, powered by solar and muscular energy. In reality, it is a vehicle family of various product lines. Four versions are available at present: transport of goods, persons, maintenance of green zones and pitchman. The one in the illustration is the model for cargo transportation. It was designed by Fabrizio Ceschin from the research unit of Design and System Innovation for Sustainability, INDACO department of the Politecnico di Milano University. A prototype was made with the support of the IPSIA Ferrari di Maranello Institute.

Example 6.12.3 MULO, hybrid quad cycle powered by solar and muscular energy, Fabrizio Ceschin

6.13.1 Prefer central electricity supply to batteries wherever possible. The batteries usually contain heavy metals (lithium, zinc, nickel, cadmium, cobalt, manganese and mercury), which have a considerable environmental impact. Among non-rechargeable batteries, the zinc-carbon-based ones are less toxic. Batteries containing lead must be phased out.

6.13.2 Applied Innovative Technologies have designed an emergency torch that uses no batteries and is powered by the movement of an internal wrapped-up magnet. The magnet slides inside a pathway installed in the handle and is moved simply by shaking the torch (30 s is enough for 20 min of light). The absence of batteries avoids employing toxic and harmful substances that are difficult to recycle.

Example 6.13.2 Shake-powered torch by Applied Innovative Technologies

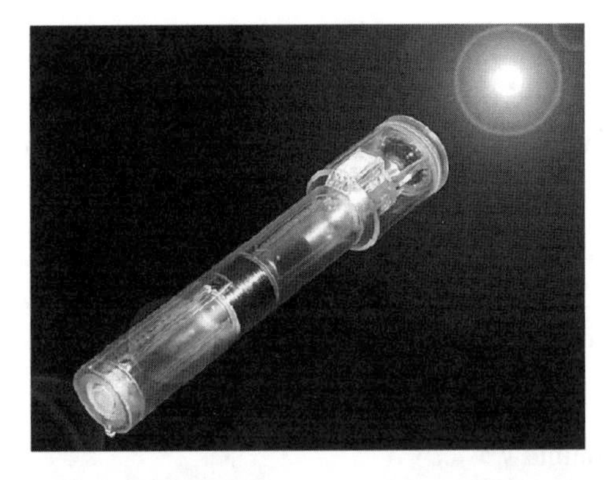

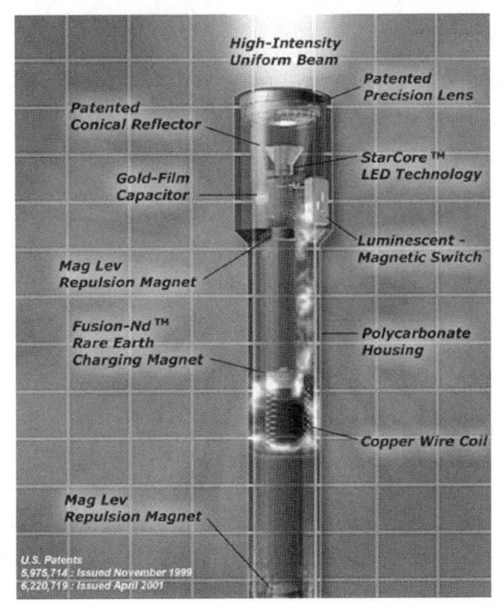

6.3 Renewable and Bio-compatible Resources

As has been already introduced, from the environmental sustainability perspective that aims to save resources for future generations as well, their renewability, or at least their non-exhaustibility, is gaining greater importance.

As before, in the interests of clear understanding and an efficient supporting structure for designers, the guidelines are divided according to the nature of the resource:

- Select renewable and bio-compatible materials
- Select renewable and bio-compatible energy resources

6.3.1 Select Renewable and Bio-compatible Materials

It has to be acknowledged that while some reserves of natural resources for producing certain materials are very limited (more limited than others), on the other hand other more renewable resources exist.

Besides, renewable substances are often biodegradable[9], which could be an advantage at the disposal stage. But only *could*; to bring the quality estimations of the biodegradable materials back into the correct environmental parameters: it is worth considering when materials originate from renewable sources and when their fast decomposition is actually an advantage, as for example packaging for humid waste.

Guidelines for selecting renewable and biocompatible materials:

- Use renewable materials (Examples 6.14)
- Avoid exhaustive materials (Examples 6.15)
- Use residual materials from production processes (Examples 6.16)
- Use retrieved components from disposed products (Examples 6.17)
- Use recycled materials, alone or combined with primary materials (Examples 6.18)
- Use biodegradable materials (Example 6.19)

Examples

6.14.1 Lixeha in Hanoi, Vietnam, has developed a bicycle prototype, where most metal parts are substituted with bamboo. All bamboo parts are treated previously to prevent cursory damage and contractions of the parts. After assembly all the bamboo parts are covered with transparent layer (glue).

The front basket, mudguards, chain guard and handles are completely made of bamboo, the pedals partly. The same material is also used to cover the frame, seat (bamboo textile), fork, handle bars and luggage rack.

However, expedience has its limits. Bamboo is definitely not strong enough to be used alone for the frame. Also, the lifespan of a bamboo bicycle is probably shorter than that of the usual steel bicycle; this material is easily damaged by humidity and is hard to maintain. Despite the limits, this prototype proves that bamboo can be used elsewhere other than just baskets and seats.

---

[9] Biodegradable polymers, for example, can be obtained from micro-organisms that feed on sugar (PHB-PHBT), or on copolymers of lactic acid, or on green corn and potato starch. Usually, these substances can be used only for short-term products.

Example 6.14.1 Bamboo bicycle by Lixeha

6.14.2 Ain Shams University, Egypt, has designed furniture made of date palm leaf midribs, which is similar in consistency to wood. As seen in the following example, this material is highly renewable.

Example 6.14.2 Furniture made of date palm leaf midribs

6.14.3 Bedouins in the Sinai have used for a long time palm leaves for sandals. Midribs separated from smaller leaves are dried and turned into sandals. No other materials are employed, not even for the joints.

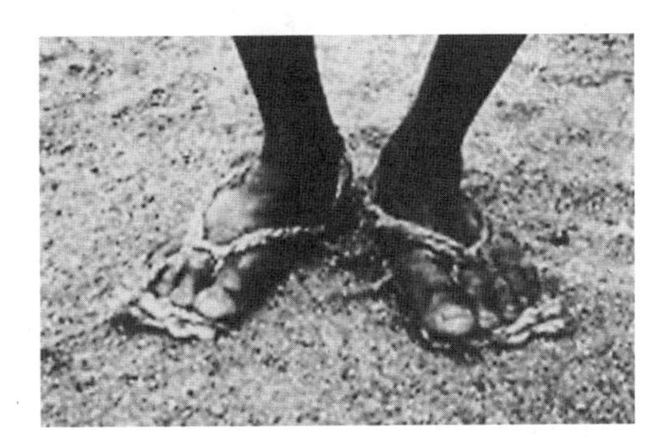

Example 6.14.3 Sandals made of date palm leaves

This material is highly renewable, because it is reproduced annually (unlike the tree-trunk). Besides, it solves leaf disposal problems for the palm tree industry. The lower leaves dry out anyway and have to be cut once a year.

6.14.4 Jan Velthuizen (Netherlands) has designed a soap container made of pumpkin. The vegetable is previously reinforced and grows inside a mould. Once matured, collected and dried, the pumpkin is carved empty to be filled with soap. Pumpkin containers are in fact traditional products.

Example 6.14.4 A soap container made of pumpkin

6.14.5 The Forest Stewardship Council (FSC) is an international certification and labelling system that guarantees that the forest products you purchase come from responsibly managed forests and verified recycled sources. Under FSC certification, forests are certified against a set of strict environmental and social standards, and fibre from certified forests is tracked all the way to the consumer through the chain of the custody certification system.

Example 6.14.5 FSC certified **a** bath massager and **b** shower brush, The Body Shop

Today, around 30 million hectares of forest in 56 countries and more than 2,200 companies are certified.

The FSC is an independent international non-profit organisation, founded by representatives from environmental groups (including Greenpeace), the timber industry, the forestry profession, aboriginal organisations and community forestry groups.

6.15.1 Do not use wood that is in danger of extinction, *e. g.* balsa, Douglas fir, ebony, Siberian larch, iroko, red cedar and teak. Not in peril of extinction are: ash tree, birch, beech tree, cherry, elm, European larch, oak, maple, pear tree, pine, poplar, black locust, Norway spruce, sycamore and walnut.

6.15.2 Metals in short supply are copper, zinc, platinum and tin.

6.16.1 In Spain a material has been developed that consists of pulverised almond shells and synthetic polymer. The almond shells are year-to-year products, thus highly renewable; besides, it reduces the waste flow of this sector of the food industry. It has also been planned to use natural polymer.

This material can be injection-moulded and in the end finished and varnished just as any other thermoplastic, combining the quality of wood and plasticity of polymers.

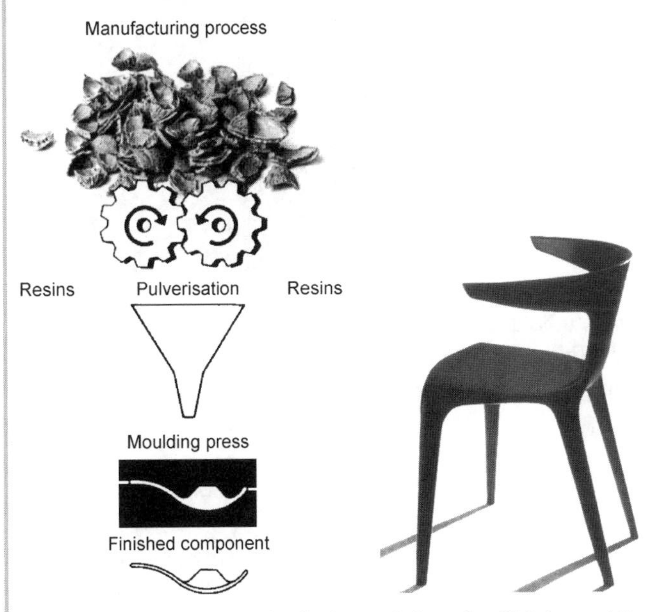

Example 6.16.1 Maderon by Gauhaus; chair made of Maderon, Alberto Lievore

6.16.2 Enea, Italy, has developed a highly pressurised hydrothermic process for lignocellulose (steam explosion). With this process rice can be transformed into material that is later used in table tops and other furniture.

Example 6.16.2 Steam explosion plant, Enel

6.16.3 Jan Velthuizen, the Netherlands, has designed tiles made of cement and mussel shells. The shells form a considerable part of the waste of the fish industry and their disposal costs can in this way be avoided.

Example 6.16.3 Tiles made of cement and mussel shells, Jan Velthuizen

6.16.4 Droog Design studio has designed packaging (Tulip box) made of dehydrated manure for tulip bulbs. The designers wanted, a little provocatively, to underline a specific problem of their country – a surplus of manure.

Example 6.16.4 Tulip Box by Droog Design

6.17.1 Ecolo is a project of Alessi, designed by Enzo Mari, a leaflet that gives instructions how to transform empty plastic bottles into vases and into furniture elements. Ecolo uses everyday discards – any used plastic bottle – thus avoiding their disposal.

Example 6.17.1 Ecolo project by Enzo Mari, Alessi

6.17.2 *The tired horse*, designed by Alan Thompson, is a horse-shaped seesaw, made of discarded tyres.

Example 6.17.2 Horse-shaped seesaw, made of scrapped tyres, Alan Thompson

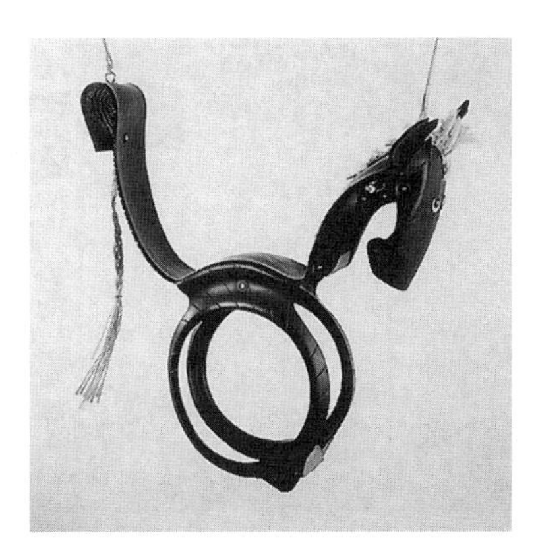

6.18.1 Philip White designed for Philips a portable CD player whose case is made of recycled CDs. Thus, even new moulds for the prototypes are not needed.

Example 6.18.1 CD player, Philip White

6.18.2 Abet Laminati, Italy, markets a plastic laminate material produced by recycling production waste – a mixture of pulverised thermosetting resins and a thermoplastic. The material is extruded into a laminated form and can be thermoformed. FIAT uses Tefor for internal panels in some of its cars.

Example 6.18.2 a–c Tefor, Abet Laminati; **d,e** Tefor is used for internal panels in FIAT cars

6.18.3 3M markets the paillette sheets, PET, which are made 100% of recycled bottles; the packaging is also made 100% of recycled paper and cardboard.

Example 6.18.3 Scotch-Brite by 3M

6.19 Mater-B is a material made of green corn starch. Because it is biodegradable, the Mater-B can be left inside the ground and does not have to be removed for composting. Optimal for planting plants.

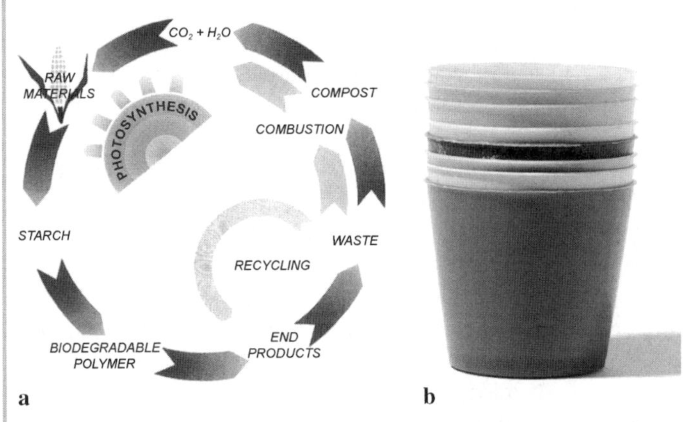

Example 6.19 **a** Production and decomposition circle of Mater-B; **b** planting pots made of Mater-B

6.3.2 Select Renewable and Bio-compatible Energy Resources

We should not forget that from the sustainability perspective the availability of energy sources for future generations is of utmost importance. Just as well we have to know which sources are exhaustible and which are renewable: for example, muscular and solar energy as well as hydroelectric and wind power.

In the end, it is preferable to engage energy transformation structures in a cascade approach that is able to capitalise on the maximum utility for mankind (Box 6.2).

Box 6.2 Energy Sources and Transformations

Transformation and Degradation of Energy

Energy has been defined in several forms: kinetic, potential, electromagnetic, thermal, chemical, nuclear, sound and light.

Different processes of energy transformation (from one form to another) have been employed by mankind to satisfy its needs: heating, artificial lighting, transportation, production of goods *etc.* Two fundamental facts are known about the transformation of energy – the I and II laws of thermodynamics.

First of all, energy does not disappear (I law of thermodynamics). It transforms and transfers, but the total amount of energy present in a closed system does not change. If it was that easy, we would not have to be worried about exhausting the energy sources, as the energy would not disappear.

Unfortunately, we know that energy forms have a hierarchical structure (II law of thermodynamics), and that the transformations occur in only one irreversible direction – the degraded direction. In more scientific terms, defining entropy as a measure to scale the energy degradation, the II law of thermodynamics says that using energy resources transforms them from one form to another, always towards greater entropy. In essence, it is impossible to reverse an energy transformation in a closed system without losing some energy: it does not disappear, but every transformation degrades it.

Mechanical and electric energy has the highest quality, in fact, they're totally convertible into work with very high returns. The lowest form of energy is heat, which in turn degrades, losing temperature.

In summary, we can imagine modern industrial society as complex machinery in which superior energy forms are employed to create a large assortment of goods and services, and are degraded into heat waste (Summers, 1971).

Actually the energy can be retrained, but only at the cost of a certain amount of energy that in turn increases the amount of entropy. Therefore and always, all transformations amount to a process of overall energy degradation.

Energy and Entropy Efficiency

Transformation efficiency from one energy form to another usually refers to the energy loss during the transformation. Meanwhile thermal efficiency indicates the ratio between achieved work (or heat) and the energy input required for it. Such calculated performance does not take into consideration the possibility of using other forms of energy for the same transformation, but with higher performance, while different transformations do have different outcomes.

It is definitely important to estimate the efficiency in relation to available alternative energy sources. Another way to analyse the performance could be defined between the useful work (or heat) produced by a certain form of energy and the maximum amount of work (or heat) that can be acquired by the best available form of energy (K.W *et al.*, 1975).

Design for higher efficiency means, therefore, preferring energy forms with a higher entropy level that are still usable (as the lowest energy forms possible).

In more general terms, we can say that it is necessary to adapt a *cascaded energy approach*. When transforming heat into mechanical energy, the maximum efficiency is achieved using the thermal energy at the highest available temperature and releasing it into the environment at the lowest available temperature. This is true in co-production, that is, in loops of energy transformations at lowering temperatures.

Guidelines for selecting renewable and bio-compatible energy resources:

- Use renewable energy resources (Examples 6.20)
- Engage a cascade approach
- Select energy resources with high second-order efficiency (Example 6.21)

Examples

6.20.1 Husqvarna (Electrolux group) has developed a solar-powered mower, equipped with an on-board computer that allows it to move independently. It works slowly, silently and continuously from dawn till dusk. It has a potency of 20 W, which is considerably smaller than that of common lawnmowers (1,000–1,500 W). It hashes the cut grass in a way that it can be later used to fertilise the terrain (reducing fertiliser consumption).

Example 6.20.1 Solar-powered mower, Husqvarna (Electrolux)

6.20.2 The Solar Shuttle is a solar-powered boat for pleasure trips, usable on lakes and rivers. Its basis is a catamaran (two parallel hulls), which supports the passenger deck. A vault of aluminium alloy protects the passengers from natural forces and hosts the solar panels. The photovoltaics convert solar energy into electricity and power the engine and on-board appliances.

Example 6.20.2 Solar-powered boat, Solar Shuttle, Kopf Umwelt und Energietechnik

6.20.3 Kone has recently presented a prototype of solar-powered elevator. The elevator runs in the aisle of solar panels that provide the power supply for the elevator.

Example 6.20.3 Experimental solar-powered elevator, Kone SuperEco

The *grid-connected* photovoltaic system (which takes a greater part of the exterior walls of the elevator aisle) converts the solar energy into electricity. In top conditions it produces more energy than needed for the elevator, and this can be saved in the local power grid and used during the night or scarce sunlight.

6.20.4 At the Institute of Technology in Lausanne, a group of chemists, conducted by Prof. Michael Graetzel, has developed transparent solar cells. The prototype is able to convert about 10% of passing sunlight into electricity.

The cells use a *redox* process similar to the photosynthesis of the plants. Two layers of glass are internally coated with titanium dioxide, which transfers the electricity to an external circuit. It is hard to estimate whether the production of these cells leaves a bigger footprint than producing normal solar cells.

But here the project has another interesting aspect: the transparency of cells allows them to be used in normal windows. In hot regions this could prove to be a further advantage: absorbing the heat would reduce the workload of air conditioners.

Example 6.20.4 Transparent solar cells, Lausanne Technology Institute

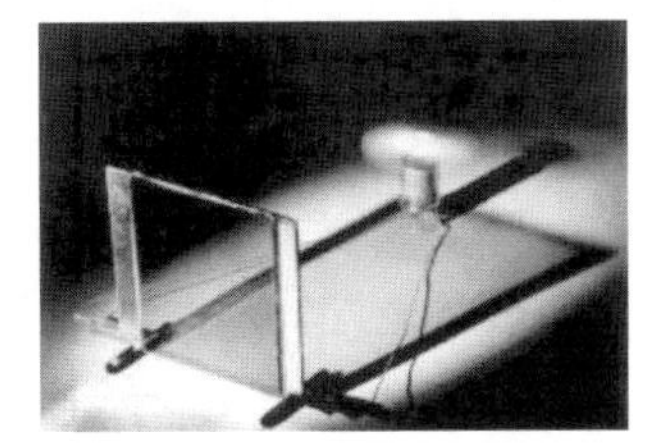

6.21 The heat pump is one of the most appropriate mechanisms for heating systems. It uses only as much energy (for the pumping mechanism) as is needed to raise the temperature from the almost endless amount of energy outside at slightly lower temperatures.

While it is a real waste to use gasoline, which gives a temperature up to 1,000°C, to heat a space to 20°C, gasoline and other fossil fuels definitely find a better use in the production of electricity.

Chapter 7
Product Lifetime Optimisation

7.1 Useful Lifetime

When speaking about environmental optimisation of a product's lifetime, two feasible strategies come to mind:

- Extending the lifetime of the product (and/or its components): designing long-performing artefacts.
- Intensifying the usage of the product (and/or its components): a design that would lead the artefacts to be exploited more frequently, minimising the time of non-usage.

Before delving into these strategies, let us recall briefly the meaning (or rather meanings) of useful lifetime.

Useful lifetime measures how long a product (Table 7.1) and its components (Table 7.2) would last under *normal working conditions*[1], maintaining its conduct and performance at accepted or even predetermined standard levels.

Measuring the useful lifetime varies from product to product according to determined functions. Some common measurements are: expected durability of the product, amount and duration of operation or the shelf-life.

Table 7.1 Useful lifetimes of different types of product

Type of product	Useful lifetime (years)
Small house appliances	3–4
Computers	2–6
Large house appliances	5–10
Cars	5–15
Electrical equipment	10–25

[1] *I.e.* properly maintained and within the prerequisite limits of mechanical stress.

Table 7.2 Durability classes of different types of timber (NEN-EN 350-2 normative)

Durability class	Durability (years) Timber protected from the moist ground	Durability (years) Timber not protected from exterior conditions	Timber
I Very durable	More than 25	50	Iroko, red cedar, teak, rosewood
II Durable	15–25	40–50	Oak chestnut, larch, meranti, mahogany, black locust
III Moderately durable	10–15	25–40	Oak, walnut tree, red American pine, Oregon pine
IV Slightly durable	5–10	15–25	Balsa, birch, elm, linden, araucaria, fir
V Non-durable	Less than 5	6–12	Ash tree, plane tree, poplar, willow tree

The end of a useful lifetime is usually called *disposal*. The main reasons that lead to product disposal are:

- Degradation of performance and structural fatigue due to normal usage
- Degradation due to environmental or chemical causes
- Damage caused by accidents or improper usage

but also:

- Technological obsolescence[2]
- Aesthetic or cultural obsolescence[3]

Let us see which guidelines there are to follow, and why in environmental terms it is meaningful to extend a product's lifetime and to intensify its use.

7.2 Why Design Long-lasting Goods?

A more perpetual product, with otherwise similar functions, will generally secure a lesser impact on the environment. While a product with a shorter life-span not only creates untimely waste, but also a new indirect impact caused by the demand for product replacement. Pre-production, production and distribution of the new product, which would cover the functions of the old one, compels further consumption of resources and creation of emissions. Figure 7.1 compares two products with the same functions, but with different life-spans, and shows in exactly which phase it is possible to avoid these impacts.

[2] Especially products in a highly innovative sector like infotechnology.

[3] Especially products of the fashion industry.

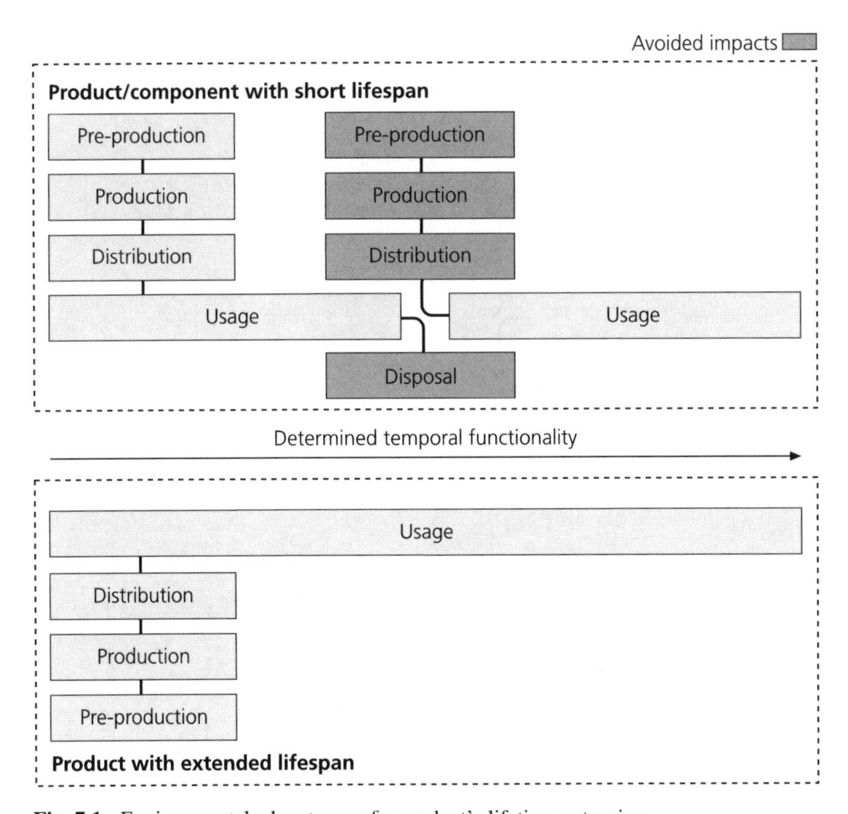

Fig. 7.1 Environmental advantages of a product's lifetime extension

With regard to the stage of usage, in reality the extension of life-span does not necessarily determine an overall reduction of the impact; on the contrary, there could be an increase if the new products are environmentally more efficient. In other words, for some products that make the greatest impact during usage, there could be an optimal length of life-span. That is, for providing the same service, technological development can offer new, environmentally more efficient products (less consumption of energy and raw materials or emission reduction), and there would come a moment, when construction, distribution and release of a new product would pay off, in terms of the environmental impact balance-sheet, due to better performance during the stage of usage. Thus, a potential limit for the length of the life-span exists, a break-even point, at which replacing the product with a new one (that provides the same service) results in less of a global impact. Or rather that impact created due to the production/ distribution of the new product and disposal of the outdated product is smaller than the reduction due to enhanced efficiency of the new product during use.

As previously stated, the main candidates for longer lasting life-span are goods that consume fewer resources (energy or materials) during utilisation[4].

[4] For example furniture or bicycles.

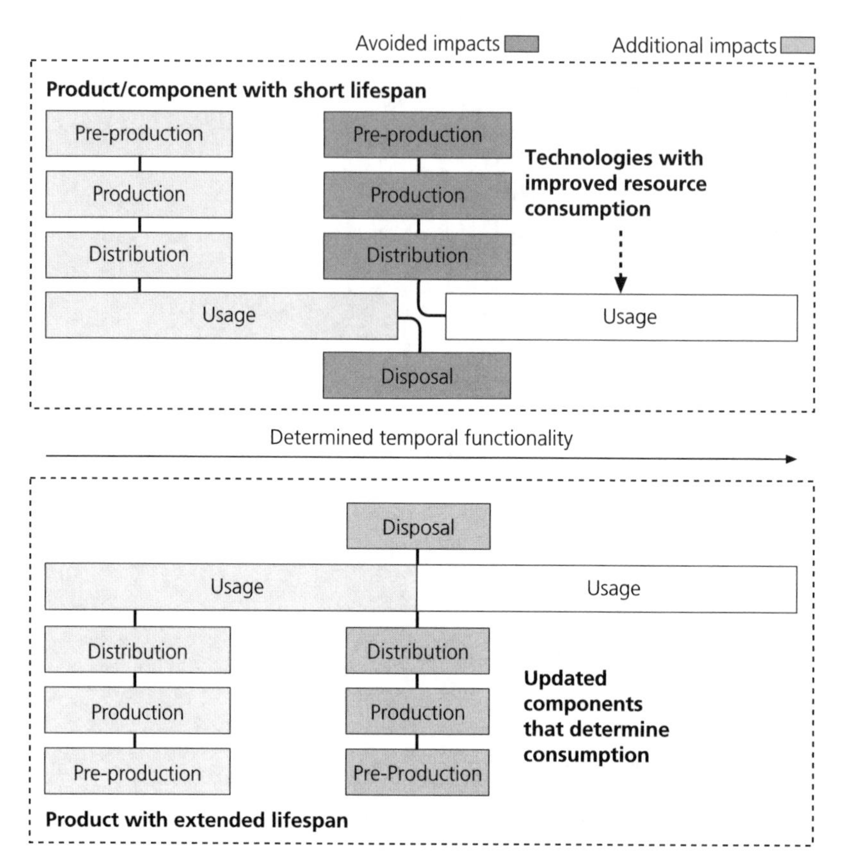

Fig. 7.2 Extension of resources consumed during a product's lifetime, by upgrading the components that are technologically obsolete

Normally, there is little sense in prolonging the lifetime of consumer goods that are destroyed during usage[5]. For other one-off goods (like packaging, newspapers, throw-away razors) the extension of their life-span is instead a prioritised strategy that should focus on product replacement with reusable items (either the whole product or parts of it[6]).

However, it is important to understand the particulars of every case. Some design interventions allow prolonging the lifetime without additional demand of resources, while in other cases the prolongation is connected with extra consumption; if this should be true, the impact of issued extra consumption should be divided by estimated extra lifetime, in order to consider and value it on the basis of time and usage. That is, the impact has to be compared with the *functional unit*.

---

[5] Actually, in some sanitary systems we can consider replacing water and detergents with mechanical washing components (for example with ultrasound).

[6] For example, replacing throw-away packages with reusable ones.

Let us look at the more critical case of products that consume large amounts of resources during usage and maintenance, for example, motor vehicles and home appliances. These problems, as stated before, cannot be solved in terms of environmental efficiency with excessively durable life-span. Actually, an interesting strategy could develop that would condition substituting only the components that determine consumption. Thus, there would be no need to pre-produce, produce, distribute and dispose the entire product, but only those parts that would decrease the overall environmental impact (Fig. 7.2).

7.3 Why Design Intensely Utilised Goods?

Any product used *more intensely* than other similar ones leads to a reduction in the actual number of those products at a given moment and place, while still answering the demand of their performance: this also determines the reduction in environmental impact. Let us clarify this concept with the help of some diagrams.

Starting with Fig. 7.3, which assumes that the lifetime is independent of the actual usage of the product. Let us imagine (in the diagram above the "usage in

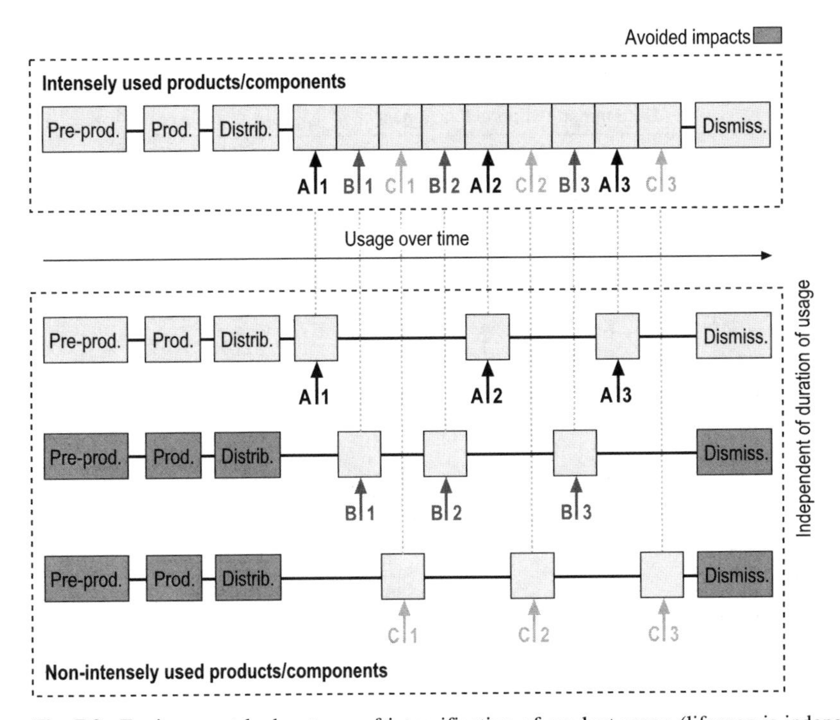

Fig. 7.3 Environmental advantages of intensification of product usage (lifespan is independent of the duration of actual usage)

time" arrow) that the product[7] was intensely used by Albert in periods A1, A2 and A3, by Beatrice during B1, B2 and B3 and by Claudio in periods C1, C2 and C3.

Now let us imagine (in the diagram under the "usage in time" arrow) another scenario, where every participant has their own product, and they use it during the same periods (reasoning with equal functional unity). Schematically, it follows that the main impact is during the pre-production, production, distribution and disposal phases of the additional products. It is true only in the case of when a product's lifetime does not depend on the time of actual usage, for example, due to obsolescence.

In other words, if products are used more intensely, their useful lifetime will pass faster without raising the global amount of products and their disposal. Thus, the more occasional the normal usage of the product and the higher its obsoles-

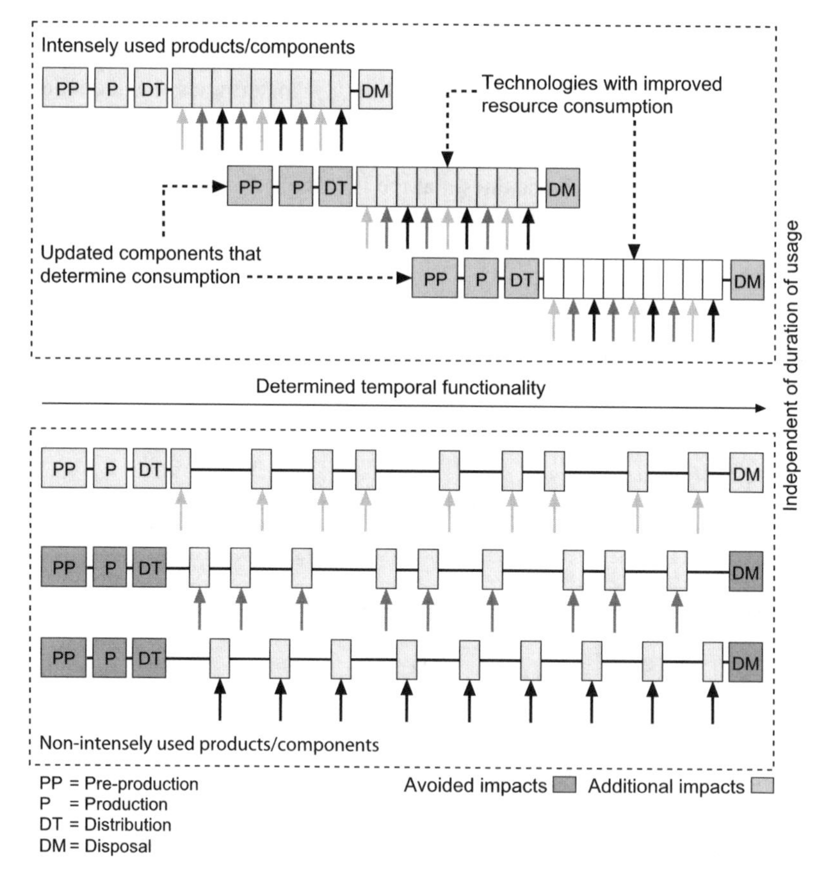

Fig. 7.4 Environmental advantages of intensification of product usage (lifespan is proportional to the duration of actual usage)

[7] For example, a shared computer.

cence (technological or aesthetic), the more the additional production can be reduced while still satisfying the same needs. Indeed, more intensive usage in general leads to a shorter absolute (time between acquisition and disposal) lifetime, but, on the other hand, it increases the time of effective usage (and reduces disposal due to obsolescence).

Let us turn to the case of when a product's durability is related (inversely proportionally) to the actual usage, meaning that more intensive usage will effectively shorten its lifespan.

Let us take the previous diagram and extend the time-line. Still reasoning with equal functionality (in these two scenarios Albert, Beatrice and Claudio use the products for the same periods of time), we must imagine the substitution of intensely used products (in Fig. 7.4 we imagine two substitutions on top of the "usage in time" line).

However, in this case the environmental advantage results solely from the potential of technological progress (greater effectiveness of the pre-production, production, usage and disposal phases) that has become available.

Therefore, one outcome lies in the potential appearance of alternative technologies (with the possibilities of reducing impacts), without increasing the number of additional products to satisfy the same needs.

Moreover, we can also take the intensification into proportional account with the quantity of goods that are produced but not sold. In other words, the smaller the excess, the greater the intensity with which we use a certain productive batch. For this purpose we have two interesting ways to reduce the excess: intervene downwards offering the products according to availability, or upwards according to demand[8].

7.4 Social and Economic Dimensions of Changes

What has been outlined so far gives an interesting framework of changes in environmental terms. But what could be the economic and social contexts within the space where design activities take place?

The general image is not very encouraging. The concept of durable and intensely used goods clashes with the tastes and values of contemporary industrialised society; in these contexts it has symptomatic value to own products (instead of using them) and to own more and newer products; nowadays, profits are connected with the quantity of products sold and well-being is measured by the growth of overall production (GDP *per capita*).

The driving force for design culture should therefore be to invent and design products that would be perceived as improvements. To *acquire value in time*, because they evoke an affection and attention, or perhaps because they are perceived as the *tools* for acquiring high quality services that they imply (disconnect-

[8] These themes are explained later as examples of related guidelines.

ing it from the value of possession). The quality of moving comfortably from one place to another, in contrast to the owning of a car (which in some European cities move with an average speed of 12 km/h).

Here, it is necessary to rethink the quality standards that are used to judge our everyday products. In design terms it is necessary to start focusing on the result provided by the fruition of the products, the actual satisfaction of needs and desires.

7.5 Optimisation Services

Is important to underline that environmental optimisation of the lifespan does not involve only the physical performance dimension of the product, but also the service component offered to the user. The topic of product durability is often connected to the existence and creation of the services of maintenance, repair and upgrading. Moreover, the intensified use should appear to the user as a service and not as a product (*e. g.* shared use products).

In the end, we can observe that for several durable products the offer of a service instead of a product could prove itself to be more eco-efficient marketing by intensifying and extending terms[9]. In this case, the producer (designer) himself will offer the services and maintain the management and property control over the services (and material products) offered. In this way, it is in his best interest to have more durable and intensely used products.

7.6 Guidelines

The guidelines for product life optimisation can be distinguished as:

- Extending the product's (and its components') lifespan
- Intensify the product's (and its components') use

Durability-related guidelines can be subdivided as:

- Designing for appropriate lifespan
- Designing for reliability
- Facilitating upgrading and adaptability
- Facilitating maintenance
- Facilitating repair
- Facilitating reuse
- Facilitating re-manufacture

---

[9] This topic is discussed in length in Chap. 10.

7.6.1 *Designing for Appropriate Lifespan*

Designing components that last considerably longer than the useful lifetime of their products often proves to create unnecessary waste. This means that once the actual life-span of the product has been identified, then its components should share a similar life-span, in order to avoid the material quality or the manufacturing processes of the components (longer durability requires greater material consumption) being greater than mandatory. Besides, many materials that improve resistance characteristics tend to create problems during disposal.

Products that are subject to fast technological evolution usually are not the best candidates for long durability: if a simple product becomes obsolete quickly there is little sense in making it last longer. More complex products that share this fate should be more adaptable[10]; the durability should be long, but appropriate.

Guidelines for designing appropriate lifespan:

- Design components with a coextensive lifespan (Example 7.1)
- Design the lifespan of replaceable components according to scheduled durability (Example 7.2)
- Enable and facilitate the separation of components that have different lifespans (Example 7.3)
- Select durable materials according to the product performance and lifespan (Example 7.4)
- Avoid selecting durable materials for temporary products or components (Example 7.5)

Examples

7.1 Kyocera laser printer drums have an equal lifespan to that of the printer itself. This is an economic advantage compared with less durable alternatives, as it removes the need for additional energy and materials to produce substituting parts.

Example 7.1 The durable Kyocera printer

[10] *Cf.* Sect. 7.6.3.

7.2 The components of computers that become technologically obsolete fast: when it is clear that they will be substituted soon, they could have shorter lifespan than components that will not be substituted.

7.3 Unlike traditional colour cartridges, which have one container for all colours, Think Tank by Canon has a different container for every colour. This solution entails a slightly larger material consumption for the cartridges, but allows the ink to be changed one colour at a time. With common cartridges, the user is compelled to substitute the whole cartridge, while it is nearly impossible for all the colours to be exhausted at the same time. This innovation lengthens the average lifespan of a cartridge and reduces the wastage of colour ink.

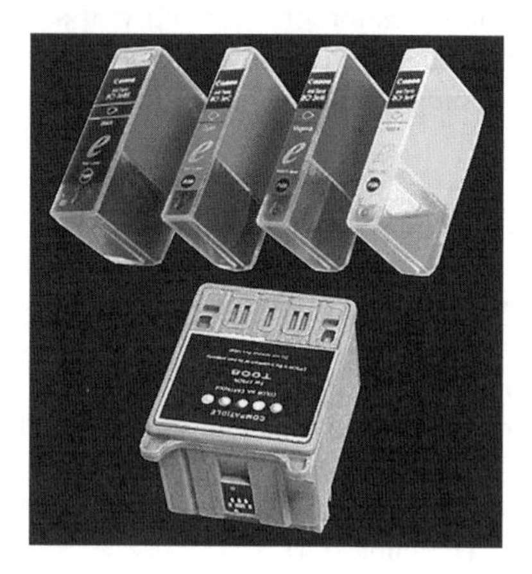

Example 7.3 The Canon inkjet printer cartridges

7.4 The Nobili faucet is equipped with a particular cartridge that reduces the wear of ceramic disks by about 50% compared with common cartridges. This almost doubles the average lifespan of the faucet compared with other products on the market.

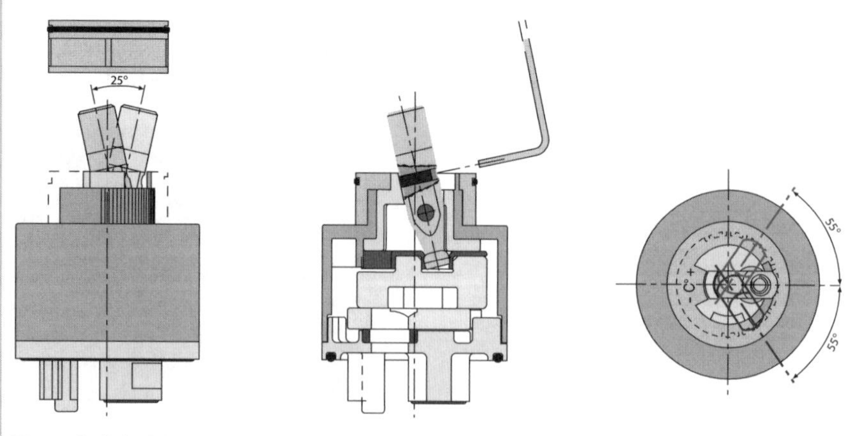

Example 7.4 The Nobili faucet cartridge

7.5 Container bags for humid waste should have a short lifespan and would preferably be biodegradable. In this case, it is no longer necessary to separate them from refuse during composting optimisation.

7.6.2 *Designing for Reliability*

Product reliability is one of the most important quality criteria. This is related together with other factors to low impact products. In fact, unreliable products, even if they are durable, will be disposed of fast. Besides, unreliable products can also pose general health and security risks.

The production of potentially unreliable goods leads to an increase in economic means and environmental impact, because they have to be repaired or substituted. As it is in both the producer's (disposal costs) and the user's (repair costs) interest to save by means of reliable goods, there should not really be any economic obstacles to this product development strategy.

Important characteristics related to the reliability are the overall amount of components, and their reliability and configuration (how different components are combined together).

Without going into detail, three general guidelines are normally required:

Guidelines for designing for reliability:

- Reduce overall number of components
- Simplify products
- Eliminate weak *links*

7.6.3 *Facilitating Upgrading and Adaptability*

Upgrading products can extend the useful lifetime because of several evolving factors and changes: technology develops, the environmental context, where the product is positioned, can change, and also the user develops physically and culturally.

Products that fast become technologically obsolete[11] can maintain their validity (remain in usage) by means of exchanging the parts that have became obsolete. This is called upgrading.

In this way, the resource consumption and waste generation are reduced in sectors characterised by short-term products and fast turn-over. In order to actually decrease the global impact a considerable part of the product (in quantitative and

[11] Computers and their hardware and software.

qualitative environmental terms) has to remain unchanged. This means that we can speak of upgrading only if a significant part of the product remains unaltered after the substitution of obsolete components. Upgrading leaves the product up-to-date with technology, but saves many components that do not have to be substituted.

As said before, adaptability is understood in relation to the changing environment in which the product could be working continuously, as well as the various stages of physical and cultural evolution of individuals. Because of this it is sound to design the products to be flexible, modular and with dynamical dimensions, performance and aesthetics.

Typical cases here are products typical to adolescents. If their performance and dimensions were flexible, then they could be used for longer periods.

Design for adaptability has to be accompanied by an adequate and innovative strategy of marketing that aims to benefit by means of guaranteed performance, to gain the goodwill of consumers attracted by the product's adaptability and by the idea that this product does not have to be replaced at short notice.

Products with built-in obsolescence that consist of great number of components are usually more approachable with regard to design for adaptability. In many cases it is preferable to design them with the goal of facilitating the removal and replacement for interchangeability of the components[12].

Guidelines for facilitating upgrading and adaptability:

- Enable and facilitate software upgrading (Example 7.6)
- Enable and facilitate hardware upgrading (Example 7.7)
- Design modular and dynamically configured products to facilitate their adaptability for changing environments (Example 7.8)
- Design multifunctional and dynamically configured products to facilitate their adaptability for the changing of individuals' cultural and physical backgrounds (Examples 7.9)
- Design products that are upgradeable and adaptable onsite
- Design complementary tools and documentation for product upgrading and adaptation

Examples

7.6 Miele offers a washing machine with intelligent system and dynamic washing program management. They can be modified via software. In this way, it is also possible to upgrade the washing programs according to the availability of new cleaning products.

---

[12] *Cf.* chapter *Facilitate disassembly* that broadens this issue and gives relative guidelines and designing options.

Example 7.6 The Miele washing machine, equipped with upgradeable software

7.7 The Mentis computer, produced by Interactive Solutions (Teltronics), is made of three components besides the screen that can be replaced independently following technological development. This characteristic allows the lifetime of the whole product to be optimised.

Example 7.7 The Mentis computer with modular functionality, produced by Interactive Solutions

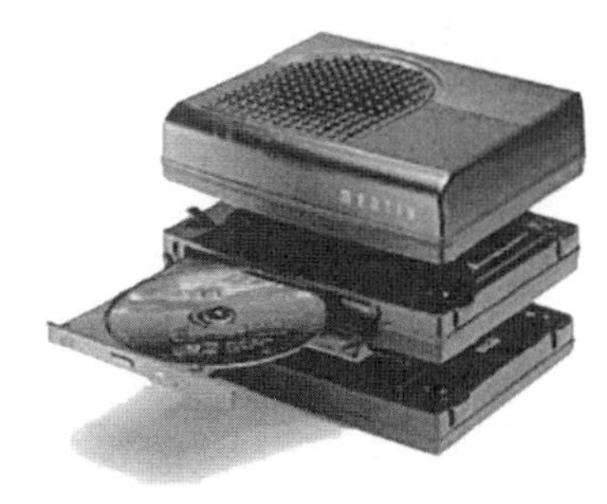

7.8 Sorin produces, like many other companies, mobile solutions for divided workspaces. The dividing walls are designed to permit the office environment to be organised with maximum flexibility.

Example 7.8 The movable dividing office walls, Sorin

The walls are also mobile to adapt according to the company's needs (and to facilitate carrying out surveys). The dividing walls are designed so that the electricity, EDP and telephone cables pass through the holes in the wainscoting, via beams attached to the ceiling or via proper shafts buried in the partition walls.

7.9.1 Leo is a cot-bed, designed by Irene Puorto, that can be spread out by lowering the elevated sides when no longer needed because of the child growing.

Example 7.9.1 The Leo bed, designed by Irene Puorto

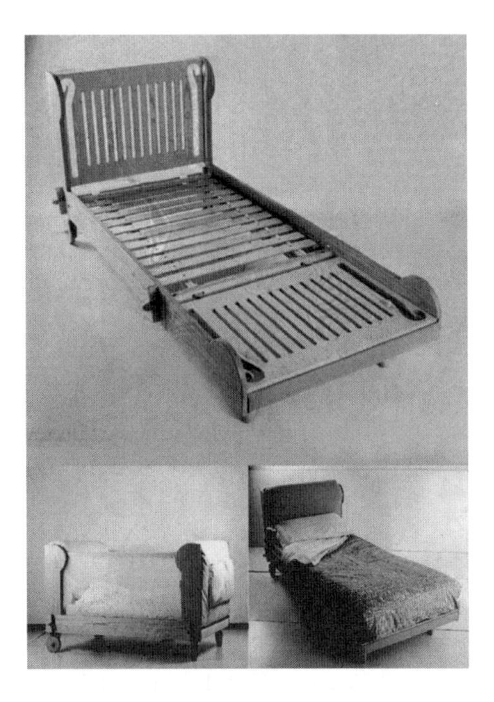

7.9.2 Back in 1979, the Klippan sofa was brought onto the market by Ikea. The sofa had easily removable covers, it was possible to buy new covers every year if a make-over was wanted, or the covers could be created oneself according to paper patterns.

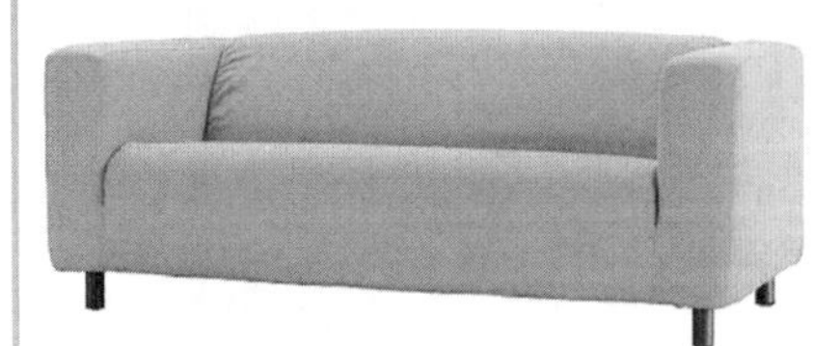

Example 7.9.2 The Klippan sofa by Ikea

7.6.4 Facilitating Maintenance

By *maintenance* is meant periodic preventive activities[13] and slight adjustments.

Appropriate maintenance can help to avoid environmental and economic costs of repairs, as well as the impact of the disposal and production entailed in replacing a product[14].

As with repairs, many complex products that are meant to be durable require proper maintenance activities that have to be facilitated by adequate design solutions[15].

Guidelines for facilitating maintenance:

- Simplify access to and disassembly of components to be maintained
- Avoid narrow slits and holes to facilitate access for cleaning
- Pre-arrange and facilitate the substitution of short-lived components
- Equip the product with easily usable tools for maintenance
- Equip products with diagnostic and/or auto-diagnostic systems for maintainable components
- Design products for easy onsite maintenance
- Design complementary maintenance tools and documentation
- Design products that need less maintenance

Examples

7.10 Nexus is a modular structure for washing machine disassembly. The structure consists of five functional subassemblies that can be disassembled into components and homogeneous materials.

Example 7.10 The structure of the Nexus washing machine, Electrolux

[13] *E.g.* changing oil or replacing the chain, tooth rim or the pinion of a motorbike.

[14] Fine tuning of a car extends its lifetime. Besides, it reduces petrol consumption (smaller costs) and emissions of toxic exhaust gases (smaller impact).

[15] *Cf.* Chap. 9.

The five subassemblies are: electronic and command components, Carboran wash drum, washer base with balance ring and pumps, and the laminated external shell (front and back wall).

7.11.1 Silver care is a toothbrush with replaceable brushes, made by Spazzolificio Piave. The small brush is fastened to the handle with a simple joint and can be easily removed by just pressing on the very end of the toothbrush itself. The handle, instead, can be used for longer, avoiding the disposal of unworn components.

Example 7.11.1 Toothbrush Silver Care with replaceable brushes, Piave

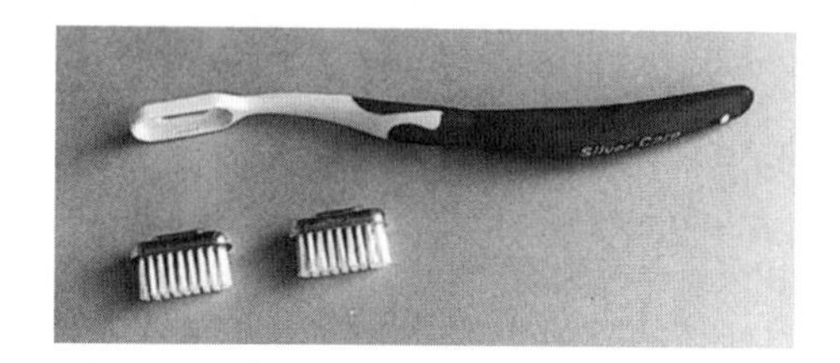

7.11.2 A common problem with mountain shoes meant for trekking and rock climbing is the uneven wear of components, which is greater on soles and smaller on the rest of the shoe.

Sportiva offers a resoling service that allows the initial quality of the shoes to be regained. Great benefits appear with the possibility of extending the product's lifetime, replacing the worn out or damaged components and retaining the others. This form of upgrading is essential considering that the sole could be reduced to being unusable in 1 year, while the rest of the shoe may last for 4–5 years. With the resoling treatment, the need to buy new medium- to long-term products is avoided and the energy and material consumption for the new pre-production, production, distribution and disposal stages are reduced. After all, if this service is managed by special centres of the same company, then the company could develop the disassembly design further, including perhaps a partial recovery of the damaged components in cascaded recycling.

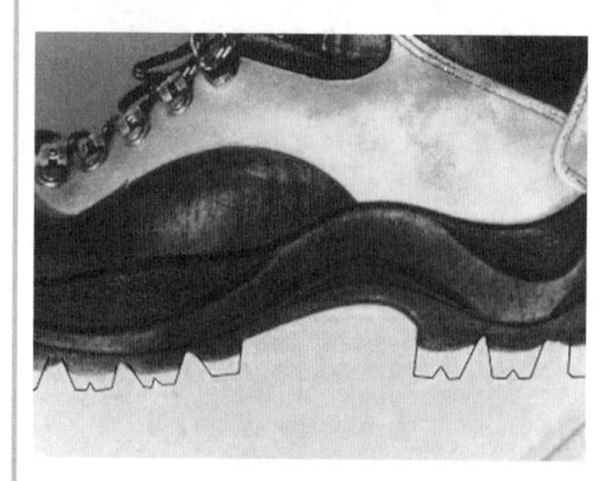

Example 7.11.2 Re-soleable mountain boots, Sportiva

7.11.3 Some shirts are sold with replaceable wristbands and collars. These are worn out faster than other parts and can be replaced, extending the lifetime of clothes and saving money for the consumer.

7.12.1 Rank Xerox has equipped its photocopiers with automatic systems for monitoring the easily worn and periodically replaceable components.

7.12.2 Some motorbikes have an oil window to check the oil level without opening the tank.

7.12.3 BMW and Rolls Royce co-produce an air engine BR 700 integrated with a monitoring system for problem identification. It has a modular structure with easy access to maintainable components to facilitate maintenance.

Example 7.12.3 BR 700 jet engine by BMW Rolls Royce

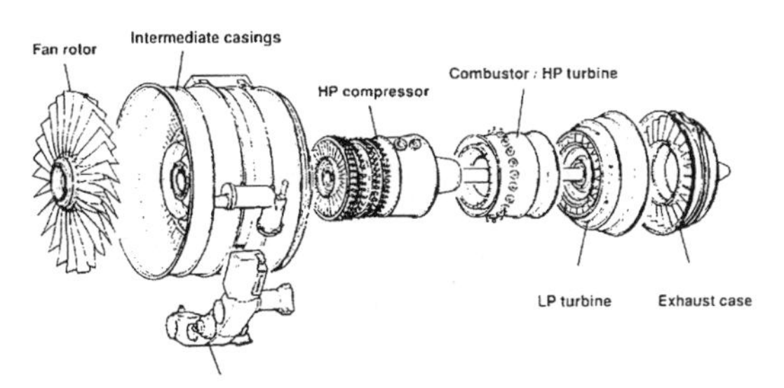

7.13.1 Piano bookshelves are exemplary in terms of long-term goods. The quality of the timber used (multilayered wood, veneered and bordered by all-wood panels) and the overall structure aim to extend the useful lifetime of the product. Different parts of the bookshelf are examined in order to offer easy assembly and disassembly: they permit easy replacement and upgrading of components and ensure longevity.

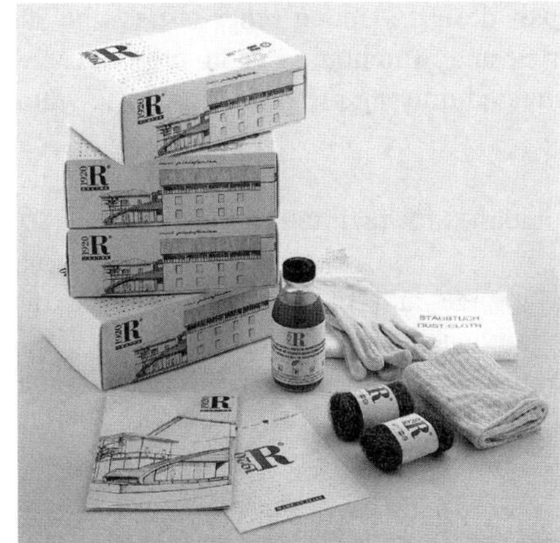

Example 7.13.1 Piano bookshelves, equipped with maintenance kit, Riva

To provide lifelong treatments, the bookshelf is sold together with a maintenance kit: a set of small instruments to independently finish (with oil-based paints or natural wax) surfaces in case they become worn out or are accidentally scratched.

7.13.2 Many cars are sold with a manual of light maintenance procedures and a few reserve parts.

7.14 The LaFutura descaler disc uses magnetic technology for water treatment in washing machines and dishwashers. It protects heating coils and mechanical parts from limestone and increases durability and efficiency.

7.6.5 Facilitating Repairs

Labour costs are the main factor that determines whether a product is eventually repaired or not. These costs are normally proportional to the complexity and difficulty of access to the repairable parts[16].

Many complex products that are meant to be durable require proper repairing activities that have to be facilitated by adequate design solutions, with a clear vision of what can and will be done and by whom. Depending on the type of the product and its usage context, the repairs could be carried out by the user or service centre; the latter can be more or less connected to the producer or distributor.

Interchangeability stands in favour of the producer of the product, who would also manufacture the spare parts[17]. Standardising, meanwhile, makes compatible spare parts and materials manufactured by different producers. Design ideas that bring along unique features or components could easily nullify the efforts of repair design. Among other issues, the special components require bigger inventories and warehouses (and more expensive) for replacement components and extended training programs for those who are responsible for the repairs.

Guidelines for facilitating reparation:

- Arrange and facilitate disassembly and reattachment of easily damageable components (Examples 7.15)
- Design components according to standards to facilitate substitution of damaged parts

---

[16] Landline telephones are exemplary of goods with a low unit cost that almost never are repaired. A new product almost always costs less than any repair procedure; because of this almost all damaged goods are replaced immediately.

[17] *Cf.* Chap. 9.

- Equip products with automatic damage diagnostics system (Example 7.16)
- Design products for facilitated on-site reparation
- Design complementary repair tools, materials and documentation (Example 7.17)

Examples

7.15.1 Interface Flooring System offers leased flooring (*e. g.* for the Southern California Gas Company). The bundle consists of: the flooring systems, installation and maintenance, as well as recovery and end-of-life treatments. The flooring system is installed in such a way that allows easy repair and replacement of damaged parts. Materials rendered unusable are recycled by the Interface Flooring System company itself.

7.15.2 Mirra chair by Herman Miller can have its damaged parts easily replaced. This has been achieved by following the rules of Design for Environment, written by Herman Miller, which concentrates on creating economic value whilst at the same time respecting the environment and involves the Design for Environment team throughout the designing process.

Example 7.15.2 Mirra chair, Hermann Miller

7.16 Precision Dell is a workstation designed to facilitate repairs (and maintenance). The box can be opened, without additional tools, by pressing in the middle of the frame; inside an instruction sheet is stuck that identifies the position of different components. Components that deal with software (mother board, RAM, video card *etc.*) are all in the back of the computer; meanwhile, hardware-related components are in the front (removable media devices and peripherals). If malfunctioning occurs the workstation informs the users via four LED lights on the front, which emit encoded messages identifying the small repairs required. In the event of greater problems, the computer is able to connect with the operator via codified audio signals, without the need for any other peripheral communication system.

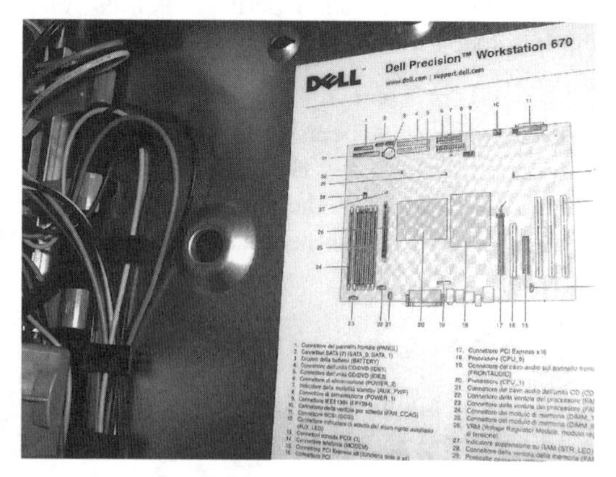

Example 7.16 Damage signal lights and component map, Precision Dell computer

7.17 Some rubber mattresses are sold with spare parts and a tube of glue for any possible repairs.

7.6.6 Facilitating Re-use

Here, re-use is understood to mean the second use of a product or its components after it has been disposed of. Repairs, cleaning and all procedures that help to save the product's integrity could be included and applied to favour the transition from one use to another. Obviously, a well-maintained product would be more easily re-used. Therefore, it is of primary importance to facilitate maintenance and repairs[18].

Products destined to be recycled have to be collected first and, without excessive operations, directed towards the same kind of usage or towards another that has smaller requirements. Necessary processes should be limited in extent and

[18] Reassess the paragraphs *Facilitate maintenance* and *Facilitate repairs*.

number, *e. g.* only cleaned or disassembled and reassembled as components of new products[19].

For some consumer goods recycling is enabled by refilling some of its parts, typically the primary packaging.

In the end, it is important, as has been repeated many times, to consider and estimate the global impact, measured against the fact that cleaning and transportation also have an impact.

Guidelines for facilitating re-use:

- Increase the resistance of easily damaged and expendable components (Example 7.18)
- Arrange and facilitate access to and removal of retrievable components
- Design modular and replaceable components (Examples 7.19)
- Design components according to standards to facilitate substitution
- Design reusable auxiliary parts (Examples 7.20)
- Design refillable and re-usable packaging (Examples 7.21)
- Design products for secondary use (Example 7.22)

Examples

7.18 One of the most common examples of re-use is the repeated use of glass bottles, which are re-collected and washed. The bottom perimeter of the bottle can be reinforced to reduce the risk of breakage due to various contacts and small occasional inevitable collisions. The transportation and cleaning activities of course have their impact, but here re-use is nevertheless expedient.

7.19.1 Rank Xerox photocopiers are disassembled after their eventual disposal, and the components go through examinations to test their eligibility for being used in new photocopiers. The photocopiers of mixed components correspond to the same quality as new photocopiers, because of the test requirements that have to be surpassed. In this system about three-quarters of the components are re-used.

[19] *Cf. Facilitate disassembly.*

Example 7.19.1 Testing and storing re-usable photocopiers, Rank Xerox

7.19.2 Grammer AG, Germany, presented in 1993 a new product line of office chairs called *Natura*. The objective of this line was to extend the lifetime of chairs and their components. *Natura* chairs are designed to last 30 years and in the end they have to be returned to Grammer, who guarantees the re-use of their parts (and materials).

The recovered goods are carefully disassembled and separated parts go through an examination. After a series of renewal operations, the components are used in the assembly of new chairs.

Grammer had to accomplish several economic tasks, the logistics of re-collection and client involvement. To ensure economic feasibility the costs are covered by a lay-away plan: the income from sales is used for the future costs of re-use.

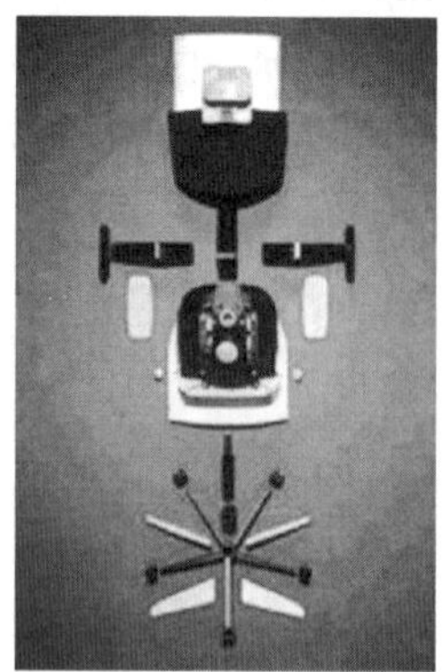

Example 7.19.2 Office chair by Grammer

7.20.1 Several filter coffee machines use re-usable filters instead of throw-away filters.

7.20.2 Several printer producers have designed and marketed rechargeable cartridges equipped with proper kits to facilitate cartridge swapping. Costs of recharging are considerably smaller than purchasing new cartridges and decrease with every printing.

7.21.1 Henkel, Germany, marketed a rechargeable glue stick, meaning that the packaging is re-usable. During the development stage extra attention was paid to facilitating the exchange and giving efficient instructions to the consumers.

Example 7.21.1 Refillable glue stick, Henkel

7.21.2 Reckit & Coleman have designed a refillable plastic container for their washing powder.

7.21.3 Peguform-Werke, Germany, employs re-usable shoe packaging; once returned, they will be re-used.

7.21.4 Supermarket chain Iper, Italy, markets its own brand of detergents, bundled with a refilling service: after the client has consumed the products purchased, he can return to the outlet with the empty container and refill it from a dispenser. The same container can be used over and over, until it is worn out and eventually recycled.

Example 7.21.4 Detergent dispenser, supermarket chain Iper

7.22 Ferrero sells Nutella in glass jars that can be later used as cups.

Example 7.22 Jar/glass of Nutella

7.6.7 Facilitating Re-manufacturing

Re-manufacturing[20] is an industrial process of renewing products frazzled by use, which will regain a condition comparable to that of new products.

[20] Besides re-manufacturing, the term *refurbishment* is used; the former refers to more consistent use of industrial operations. In this book we will not use an extra subdivision, though we see that the real difference is made by using or not using such operations for component modification.

Industrial equipment and high-cost products[21] that do not have to be changed often, are, economically speaking, more suited to re-manufacturing.

Is important to design for the recovery of disposed products, with marketing campaigns or re-collection logistics. Ultimately, it is essential to provide an adequate infrastructure for storage and inventory.

To facilitate re-manufacturing, it is of paramount importance to have favourable removal, replacement and interchangeability processes for the components of the entire product line (*cf.* Chap. 9).

Guidelines for facilitating re-manufacturing:

- Design and facilitate removal and substitution of easily expendable components
- Design structural parts that can be easily separated from external/visible ones (Example 7.23)
- Provide easier access to components to be re-manufactured
- Calculate accurate tolerance parameters for easily expendable connections
- Design for excessive use of materials in places more subject to deterioration
- Design for excessive use of material for easily deteriorating surfaces (Example 7.24)

Examples

7.23 Wilkhahn, Germany, is a producer of office furniture. They have developed new product lines that follow the guideline of separating structural components from external elements. This was related to the objective of recovering discarded furniture and reselling after re-manufacturing.

7.24 For the cylinders of several motorbike models pistons with an enlarged diameter are available. It is possible to remove the light ruling from the interior cylinder walls with an adjusting operation that removes a small part of the materials. In the end, only pistons (with an enlarged diameter) have to be replaced, not the entire engine.

7.6.8 Intensifying Use

Designing for the intensified use of products (and/or their components) entails design orientation towards multifunctional products and common replaceable components or products for integrated function.

---

[21] *E.g.* plane and bus engines, production equipment and office furniture.

Even more remarkable is the design for products for shared use. Collectively used products that offer simultaneously the same performance for many users are significantly more efficient. Moreover, the intensification of production batches is achieved by reducing the surplus. This can be attained by offering products on demand or on availability.

Guidelines for intensifying use:

- Design products and services for shared use (Example 7.25)
- Design multifunctional products equipped with replaceable common components (Examples 7.26)
- Design products with integrated functions (Examples 7.27)
- Design products or components on demand (Example 7.28)
- Design products or components on availability (Example 7.29)

Examples

7.25 Stattauto in Berlin is one of the oldest car-sharing networks. This kind of service consists of a certain amount of cars available to members. Members pay a service fee and a rate per kilometre. Every user has to book in advance his car usage, but from the other side the costs are considerably lower and a wider variety of cars are available: economy, luxury and transportation cars.

7.26.1 Several appliances combine telephone, fax and answering machine in one product.

7.26.2 Laptop computers combine the keyboard, screen and hard-drive into one product that has smaller dimensions.

7.27.1 Several kitchen appliances are equipped with different components to mince, chop, knead and mix food products. Only one engine and base of unit is needed, which reduces the total amount of components with equal performance.

7.27.2 A wristwatch, produced by Swatch, besides its traditional function, turns into a lift ticket in skiing areas.

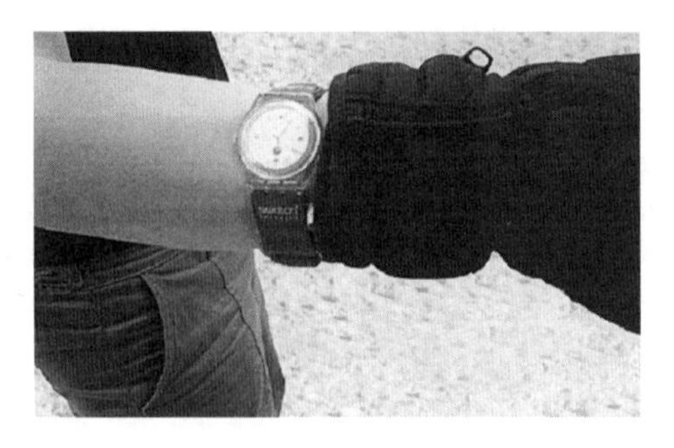

Example 7.27.2 Wristwatch – skiing lift ticket, Swatch

A daily ticket can be loaded into the internal chip at the usual ticket offices. To access the service on skiing paths the watch has to be put close to the sensors of the elevators. The integration of lift ticket with a normal wristwatch reduces the employment of different materials: it is no longer necessary to manufacture physical tickets, their carriers, rubber bands *etc.*

7.28 Lampi di Stampa, a company of the Gruppo Messagerie Italiane, is an Italian publishing company that offers print-on-demand services. Thank to digital technologies, the company can print books when ordered, even if just one copy is needed.

In the print-on-demand system, the books initially exist only in a digital format, but are listed in official catalogues. When someone wants to buy a copy, he orders it in a bookshop or via the internet; the book will be printed and sent directly to the purchaser within a few days. The price is comparable to or lower than the price of a normally printed book.

With print-on-demand only the amount of books actually purchased are printed, avoiding the storage surplus and eventual pulping of the unsold books. The system also has other economic advantages related to the dematerialisation within the production and distribution processes.

7.29 Odin Holland brings organic food directly to subscribed consumers. The consumer receives the produce by paying a fixed subscription fee, and on weekly basis a bag with assorted fruit and vegetables are sent to the consumer, enough for about 4 days.

The clients do not specify the mix they receive; the selection is made by Odin according to the products available, often from the same region in which the consumers live.

Example 7.29 Ecological food and vegetable delivery – on-availability service, Odin

Odin has a collaborative network of about 100 independent producers, who cultivate approximately 500 hectares of land, and they have a direct contract that excludes intermediaries.

About 28,000 families have subscribed to Odin.

This service allows to discards to be reduced (the surplus quantities produced for unstable demand), because the choice of consumer goods is made according to seasonal production.

Chapter 8
Extending the Lifespan of Materials

8.1 Introduction

Extending the lifespan of materials means making them last longer than the products they are part of. This kind of *reincarnation* of materials can take place through two fundamental processes. The materials can be re-processed into *secondary raw materials*, or combusted (incinerated) to recover their energy content.

The first case – re-processing of *secondary raw materials* – is called *recycling*, if these materials are used to manufacture new industrial products[1]; or *composting* if the *secondary raw materials* are transformed into organic matter and used as natural fertiliser, usually, mixing and soaking soil with decayed waste (also called *humid waste*).

In any case, the environmental advantage is doubled (Fig. 8.1). First of all, it prevents the environmental impact of materials in a landfill; second, pre-production of materials or energy has access to non-primary raw materials, thereby preventing the environmental impact of producing a similar amount of material or energy from primary raw materials. The impact of processes that *does not occur* is considered an indirect environmental advantage.

It is a common understanding that recycling is one of the best solutions for a great variety of environmental problems; thus, it is necessary to underline the fact that recycling processes also have their impact; among others they require transportation. To calculate correctly the global environmental effect of recycling strategies, one has to deduct the impact of the process from the above-mentioned indirect advantages.

Experience and several calculations show, however, that recycling materials is usually environmentally profitable (Fig. 8.2). This is why it is recommended to design products that are easily recyclable, but with the analysis of the entire life

[1] *Cf.* Box 8.1.

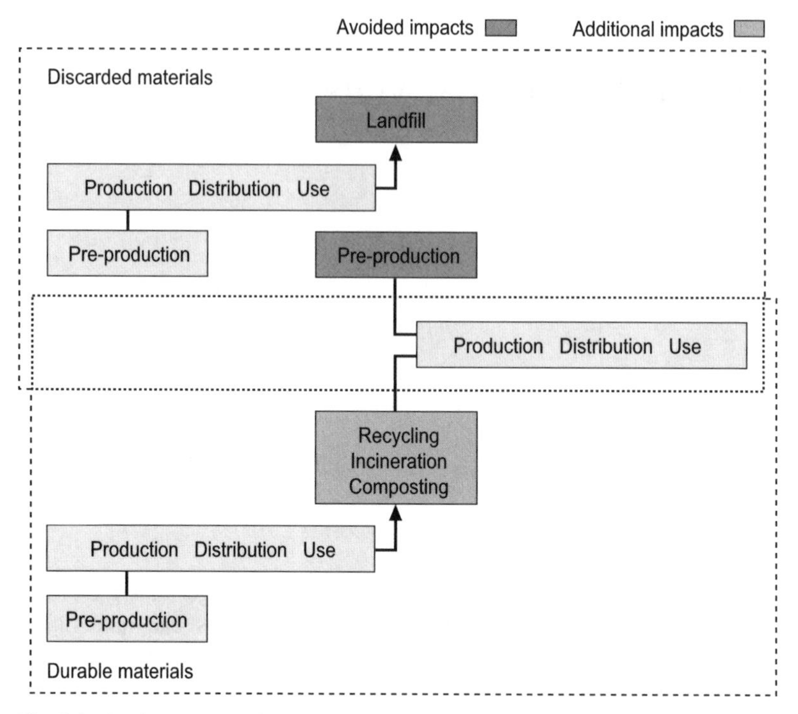

Fig. 8.1 Environmental advantages of extending the lifespan of materials

cycle in the background; neither the impact of the recycling process nor the consistency of the hypothesis that it will really be recycled should not be neglected.

In the end, it could be hard to judge whether the final impact will be higher in the case of combusting or recycling. While it is true that burning plastics, paper and carton emanates smoke[2] and leaves behind slag[3], it is just as true nowadays that the incinerators have advanced filtering systems due to strict legislation defending the environment and health. Often the high costs of particular treatment plants determine the greatest hindering of their proliferation. Or the difficulties can be for political and social reasons; one just has to think about the frequent protest demonstrations of people who do not want to have an incinerator in their back yard.

Also taking into consideration a careful analysis that would indicate more precisely the advantages and disadvantages of incineration and recycling, we can however formulate an environmental priority, repeating what has been told about the cascading approach: to proceed with recycling until the degradation of the characteristics of recycled material makes it impossible to use any longer; and from that point we can start thinking about combusting it for its energy content.

---

[2] Smoke containing chlorine and bromine is most aggressive. Dioxins emanate from the combinations of these and other organic residuals.

[3] From this waste nothing but few metals can be recovered.

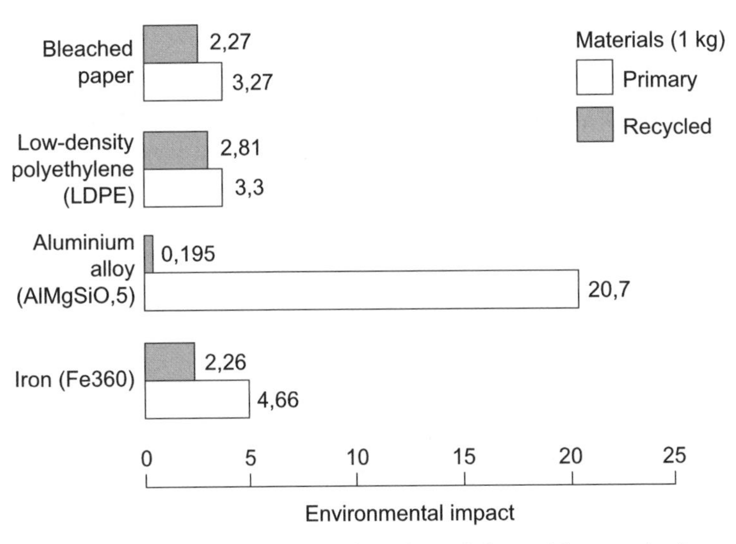

Fig. 8.2 Compare the impacts of virgin and recycled material pre-production

These considerations do not necessarily count with all kinds of discarded products. It is definitely more thoughtful to incinerate materials that contain toxic organic compounds.

In the end, one has to obviously bear in mind some characteristics of the recycling, composting and combusting processes in order to secure efficiently carried out material lifetime extension.

Recycling plays a notable role in designing for the low environmental impact of materials' end-of-life treatments. For this reason we will describe in detail this process (which materials to recycle, the routes and stages, as well as the infrastructure that would allow it to function) in Box 8.1.

Box 8.1 Recycling

Classification of Recycled Materials

First of all, the materials are classified according to the stage of the life cycle at which they become available for recycling. Here, two fundamental classes can be distinguished: *pre-consumer* and *post-consumer* materials.

Pre-consumer materials in turn can be divided into the following categories:

- Discarded materials and by-products of a certain production stage that are usually recycled within the same process.
- Materials derived from the discards, waste or surplus that are generated at any production stage, but at the original production stage.

Pre-consumer materials are generally clean, well identifiable and adaptable for high-quality recycling.

Post-consumer materials derive from products and packaging after being dismissed by the consumer. Usually of low quality and generally more difficult to recycle.

Recycling Routes

There are two fundamental routes for recycling: closed- and open-loop.

Closed-loop Recycling

Closed-loop implies a system whereby recovered materials are used instead of primary raw materials in the same product or components that they were used for in the first place. Theoretically, closed-loop recycling can be self-contained for a determined period of time, without requiring additional primary raw resources.

In reality, we cannot speak about a totally closed loop. Given the energy required by the recycling process, it is in fact hard to imagine a completely self-contained system.

Besides, the process degrades itself, because every recycling course creates its own discards and usually the characteristics of recycled materials reduce over time, as is the case with polymers. Meanwhile, for chemical/physical reasons metals recover more easily their primary characteristics, but also, because the industry of recycling metals was created long before that of polymer recycling. Because of this we can still expect improvements in polymer recycling, especially considering the perspectives of chemical recycling.

Post-consumer materials are generally difficult to engage into closed loop recycling. This model has often been adapted in industrial contexts.

Although we cannot speak about a totally closed-loop, a system that returns to the producer the materials of its products has several advantages.

- It impels the producer to design to facilitate recycling, thanks to the personal interests involved with this kind of operation.
- It facilitates the treatment of disposed products; usually, the producer knows better the configuration of the products being recycled. A third party recycler may have more problems, with, for example, identifying materials and separating them.

The disadvantage, meanwhile, lies in the limited amount of simultaneously operated materials and products. For small entrepreneurs the disadvantage is more serious: such a system might be economically unfeasible for them.

Open-loop Recycling

This term implies material recycling by producers than other than those of the original products. Usually, it is concerned with post-consumer materials.

The advantages are that:

- Operations entail high quantities of materials that allow economies of scale.
- There is a wider range of options for secondary usage of recycled materials.

Disadvantage lies in the fact that one has to deal with a mixture of products that come from different producers.

As it would not be possible to know exactly where the materials had come from and therefore also their characteristics, normally these materials end up in products' components with inferior requirements than the actual characteristics would allow. Thus, the recycled materials are actually undervalued.

Stages of Recycling

With regard to post-consumer materials the following stages can be distinguished:

- Collection and transportation
- Identification and separation
- Disassembly and/or crushing
- Cleaning and/or washing
- Pre-production of secondary raw materials

Collection and Transportation

Discarded products have to be first of all re-collected and transported to recycling sites or warehouses. The collection stage can involve many different participants, including end-users. Then again, the consumer assumes a fundamental role, being the one, who confirms the end of a product's life, and who starts the recycling cycle. In fact, this stage often decides if there are enough economic and environmental advantages in recycling or not.

The operations of collection and transportation should not be underestimated, not in the terms of logistic planning nor with regard to environmental impact. In fact, this is stage often the first to judge if recycling is economically or environmentally feasible.

Identification and Separation

When unwanted products arrive at the recycling site one has to identify the materials and decide what to recycle and how, and what to discard.

Disassembly and/or Crushing

A material has to be separated from others with which it is incompatible in order to be recycled. This does not just mean separating plastic from steel, but also some thermoplastics from other types of thermoplastics. In fact, most thermoplastics have different melting temperatures and cannot be recycled together.

At the same time, some materials can be recycled together without affecting the recycling mix. Some combinations of materials may even have synergic effects. Materials that can be recycled together without compromising the recycling results are said to be compatible with each other. The compatibility of two materials is actually measured by the quality of the resulting mixture. The characteristics of the latter depends on both the quality of the original materials and the homology between the two stages.

The separation of the materials can be achieved through disassembly or via crushing and following separation[4].

Cleaning and/or Washing

Various materials, even after separation[5], can still be contaminated[6], in which case they undergo cleaning in order to avoid compromising the characteristics of the recycled results.

Pre-production of Secondary Raw Materials

A material can be re-employed straight away, though usually they are re-processed and upgraded by specific processes and complemented with additives.

The Progression of Recycling Costs

The total cost of recycling and its economical feasibility is defined by a series of variables.

- The costs of collection, transportation and storage[7].
- The costs of disassembly and crushing: easily disassembled materials reduce the time and costs required for separating materials.
- The costs of primary raw materials: the natural resources for various materials are becoming exhausted; therefore, their costs are rising. Such developments show the advantage of recycling.
- The costs of landfill: the decreasing availability of appropriate spaces and legislative action on the requirements of landfill sites increase the discharging costs.
- The price of recycled materials: easily disassembled materials are usually less corrupted; cleaner materials are better at retaining their characteristics when recycled and therefore their market value.

Here it is clearly shown that a moment where the designer can step in is at the disassembly of the product. This will be defined in detail later, together with possible options[8] (Example 8.1).

---

[4] For different options for separating, *cf.* Chap. 9.

[5] In the case of plastics that are cleaned before identification and separation.

[6] *E.g.* the labels and adhesives stuck on the products.

[7] These kinds of cost are especially present when dealing with the collection of materials that are going to be incinerated, composted or otherwise disposed of.

[8] *Cf.* Chap. 9.

Examples

8.1 Recycling plastic water bottles – describing the usual characteristics of the materials and possible stages at which the materials can be recycled.

Materials:

- PVC (permeable with CO_2)
- PET (impermeable with CO_2)

Possible recycling systems include the following stages:

Pre-consumer:

- During production

Post-consumer:

- Collection
- Separation
- Grinding
- Cleaning (second separation)
- Granulating
- Recycling

Available methods for separating such plastics:

Physical identification: the separation can be handled manually (with visual identification) or automatically (using scanners or comparing against models). Human errors and deformed products can bring along non-accurate results.

Chemical scanning: identifying components by scanning the molecular composition of polymers. Employed with PVC containers that are easily identifiable with chlorine. Recently, infrared analysis has also widened this area of conduct.

Separation according to density: employed after the grinding and exploits the different densities of different plastics.

Electrostatic selection: another innovative method that uses electrostatic charges on granulated polymers.

8.2 Guidelines

Designing for extended material lifespan can be broken down into the following guidelines[9]. Regarding the recycling, these are essentially the same for all recycling stages.

- Adopt the cascade approach
- Select materials with the most efficient recycling technologies
- Facilitate end-of-life collection and transportation
- Identify materials

[9] When speaking of post-consumer extensions.

- Minimise the number of different incompatible materials[10]
- Facilitate cleaning
- Facilitate composting
- Facilitate combustion

8.2.1 Adopting the Cascade Approach

Adopting a cascade approach means pre-planning and designing the route of the recycled materials from one product or component to another, each with lower requirements (Fig. 8.3).

This strategy is based on the presupposition that in comparison with primary raw materials, the recycled materials will inevitably lose some of their characteristics, either for economic or technological reasons. Therefore, we can imagine a route for cyclical recycling, where materials are employed for components with gradually lower requirements (Example 8.2).

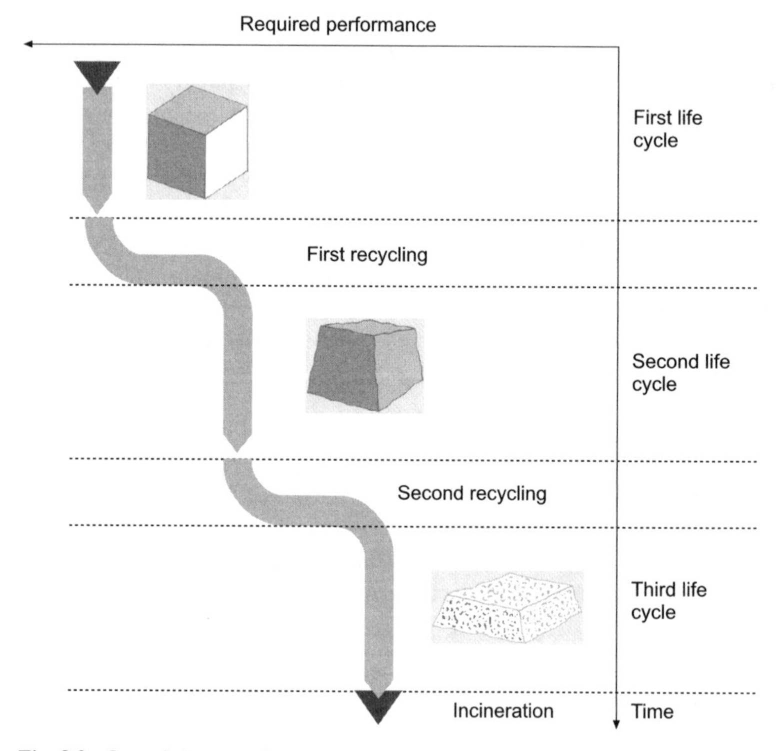

Fig. 8.3 Cascaded approach

[10] Chap. 9 returns to this topic.

When finally the characteristics no longer satisfy any requirements, the materials can be combusted in order to recover their energy content.

Guidelines for implementing a cascade approach:

- Arrange and facilitate recycling of materials in components with lower mechanical requirements (Example 8.2)
- Arrange and facilitate recycling of materials in components with lower aesthetical requirements (Example 8.2)
- Arrange and facilitate energy recovery from materials throughout combustion

Example

8.2 FIAT has developed a so-called FARE (Fiat Auto REcycling) system, a circuit of cascaded recycling that engages some car components that were not recovered before, *e. g.* after the first life cycle the bumpers made of PP are used for air conveyors, and because they are not visible they also have fewer aesthetic requirements. Recycled polymers, in fact, have barely controllable colours and usually become darker. After the second life cycle as conveyors, they are recycled and employed as car mats. It is worth mentioning that this system has also involved for efficient functioning other participants, starting with allied industries (the producers of the components) and finishing with the final recyclers.

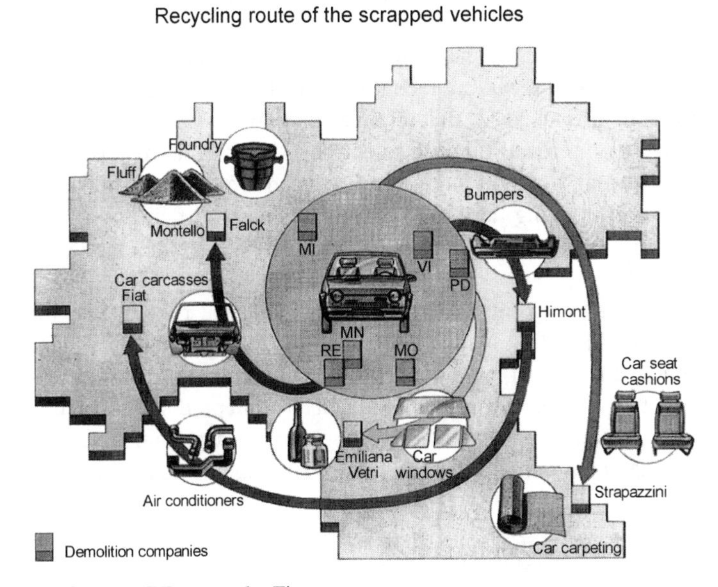

Example 8.2 a FARE system by Fiat

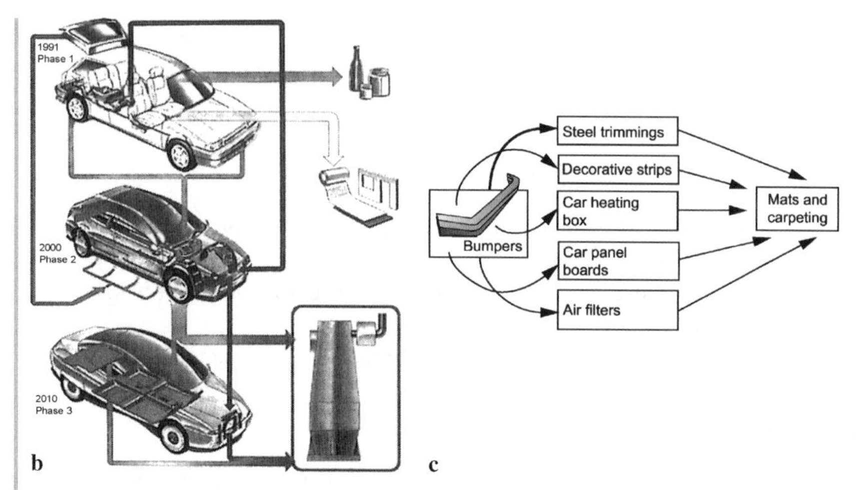

Example 8.2 **b** Cascaded approach of the FARE system, **c** Cascaded recycling of the bumpers in the FARE system

8.2.2 Selecting Materials with the Most Efficient Recycling Technologies

Not all materials are recyclable in the same way, or rather, the performance characteristics of the recycled materials can deviate to some extent from the characteristics of the primary materials. This also depends on the specific material and its cleanness during the recycling.

Selecting easily recyclable materials does not entail just identifying materials predisposed for a certain recycling technology (technologies that exist or are under development), but also materials that retain a competitive value once recycled[11].

Besides, one has to recall that the selection of recyclable materials should not be disconnected from a careful analysis of their environmental impact throughout their life cycle. Employing an easily recyclable material that has a high energy intensity during pre-production might not be worth it, either economically or environmentally[12].

After a series of treatments, due to chemical/physical reasons, the metals and glass are almost completely recyclable and retain their original performance (recovery would be absolute, if it were not for the contaminants).

Thermoplastic polymers are thermally recyclable, but partly lose their characteristics, either during usage (mechanical stress, chemical degradation, environmental degradation *etc.*) or recycling processes (after grinding or granulating; Table 8.1).

---

[11] Materials with high market values after recycling are, for example, chrome, nickel, aluminium, steel, ABS, ASA, PA, PC, PMMA, POM and the thermoplastic elastomers.

[12] A careful impact analysis (as we are going to see in Chap. 12, the Life Cycle Assessment is one of the most secure methodologies) can identify the most feasible hypothesis.

Thus, the thermoplastic polymers are recyclable, but the characteristics of recycled materials (that actually also depend on the cleanness of recyclable material) are inferior to those of primary raw materials.

In this regard, it is better to recycle the thermoplastics straight after grinding (and possible washing) without undergoing the granulating processes.

Thermosetting polymers are not thermally recyclable, but could be crumbled and recycled into extenders in other thermoplastics up to 30% of the total mass. Therefore, they are less recyclable than thermoplastics. In future, the thermosets will probably be able to be recycled chemically, but today these processes are still too expensive (Table 8.1).

In the end, it is better to avoid additives that reduce the performance of recycled materials.

Table 8.1 Typical thermoplastic and thermosetting polymers

Polymers	Thermoplastics	Thermosettings
Plastics	High-density polyethylene (HDPE)	Thermosetting polyesters
	Low-density polyethylene (LDPE)	Unsaturated polyesters (UP)
	Polypropylene (PP)	Carbamide
	Polyvinyl chloride (PVC)	Epoxide resins
	Polystyrene (PS)	Melanin resins
	Shockproof polystyrene (HIPS)	Phenolic resins
	Acrylonitrile butadiene styrene (APS)	
	Acrylonitrile styrene (SAN)	
	Polymethyl methacrylate (PMMA)	
	Aliphatic polyamide (nylon, PA)	
	Aromatic polyamide (Kevlar, PA)	
	Polycarbonate (PC)	
	Polyethylene terephthalate (PET)	
	Polybutylene terephthalate (PBT)	
	Polyoxymethylene (POM)	
Elastomers	Ethylene propylene (EPDM)	Styrene butadiene (SB)
	Styrene butadiene styrene (SBS)	Silicone rubber (SR)
	Styrene isoprene styrene (SIS)	Isobutene isoprene rubber (IIR)
	Styrene ethylene butadiene styrene (SEBS)	Acrylonitrile butadiene rubber (NBR)
	Thermoplastic urethane (TPU)	Thermosetting polyurethane (PU)
	Thermoplastic polyester (SPE)	Polyisoprene (natural rubber) (NR)

Guidelines for selecting materials with the most efficient recycling technologies:

- Select materials that easily recover after recycling the original performance characteristics (Example 8.3)
- Avoid composite materials or, when necessary, choose easily recyclable ones
- Engage geometrical solutions like ribbing to increase polymer stiffness instead of reinforcing fibres (Example 8.4)
- Prefer thermoplastic polymers to thermosetting
- Prefer heat-proof thermoplastic polymers to fireproof additives (Example 8.5)
- Design taking into consideration the secondary use of the materials once recycled (Example 8.6)

Examples

8.3 Alpha Metals from the Cookson group have developed a withdrawal system for the packaging of their welding paste. Packaging is made only of steel and after withdrawal can be completely recycled (together with the remains of the paste) in the same production process.

8.4 Glass, nylon and carbon fibres are used in polymer components for improving their rigidity. The components can still be recycled, but during the process fibres are shattered and therefore the polymers lose part of their rigidity.

8.5 Completely recyclable thermoplastics that do not require fire-retardants, like PS, PPO and PC, can easily replace fire-retardant additives that contain heavy metals or components on the PBB and PBDE basis.

8.6 To predispose the recyclability of PET bottles for fibre production, coloured materials should be avoided because they compromise the characteristics of the material.

8.2.3 Facilitating End-of-life Collection and Transportation

In order to ensure that a product designed for recycling would actually enter into the recycling system at the end of its life, the route of recycling and the engagement of secondary materials have to be carefully planned.

For such a system to function properly its economic and technological feasibility has to be assessed first. Here, two schematic situations can be presented[13].

Either entirely or partly an already existing collection system can be employed that is run by someone else[14], or rather such a system can be designed and brought

[13] *Cf.* Box 8.1

[14] Such as public bodies or contractors, who deal with the end-of-life collection of certain products and materials.

to life from scratch. Clearly, the economic and organisational/strategic implications, not to mention the consequences of design choices, change considerably in these two cases. In brief, it has to be verified whether such recovery systems are already available or if it is possible to introduce and design one.

Guidelines for facilitating end-of-life collection and transportation:

- Design in compliance with a product retrieval system (Examples 8.7)
- Minimise overall weight
- Minimise cluttering and improve stackability of disposed products
- Design for discarded products' compressibility (Example 8.8)
- Provide user with information about the disposing modalities of the product or its parts (Example 8.8)

Examples

8.7.1 If we are going to rely on an available system of differentiated waste collection, then we have to know exactly which kind of waste they are dealing with. It would not make much sense to design easily recyclable polymers in areas in which the recycling of polymers is not provided for.

8.7.2 By law, the lead and sulphur acid of car batteries has to be re-collected; in this regard, it would be interesting to design for the recyclability of polymer parts as well, making use of the existing re-collection system.

8.8 Evian has designed a spiral-ribbed plastic bottle that facilitates the flattening of empty packaging and reduces the transportation and collection space (increasing the density). The consumer is the one who is supposed to flatten the bottle, so clear guidelines (pictures) are given on the bottle.

Example 8.8 Water bottle, Evian

8.2.4 *Identifying Materials*

To facilitate material sorting for recycling, it is of utmost importance to designate constituent materials, especially in the case of complex products or of materials whose recycling route is not strictly standardised[15] (Fig. 8.4).

Fig. 8.4 Common codification on plastic materials

Guidelines for material identification:

- Codify different materials to facilitate their identification
- Provide additional information about the material's age, number of times recycled and additives used

Examples

8.9 Arrange proper designations for facilitating the correct disposal of batteries.

8.10 In the case of plastic bottles (or any objects really that have a proper base to stand on) it is advisable to designate on the bottom.

8.11 Example of moulded codification inside the bumper of a car by FIAT, in compliance with the FARE project.

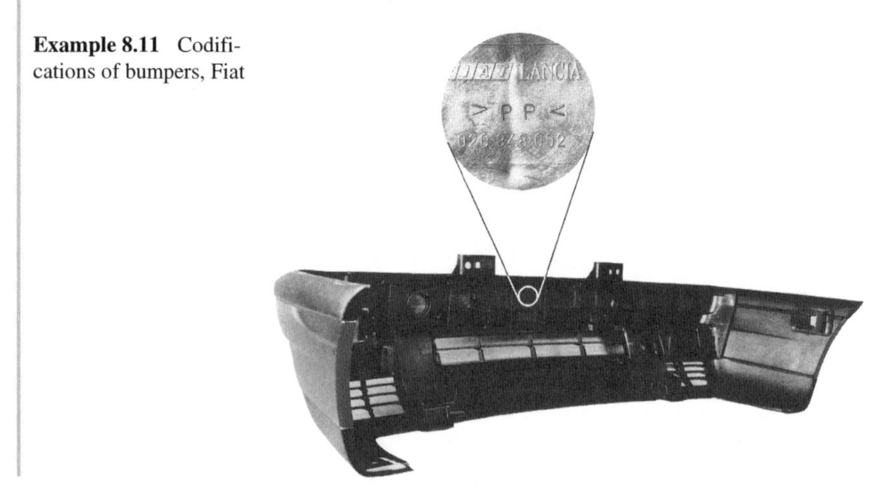

Example 8.11 Codifications of bumpers, Fiat

[15] *E.g.* in UN 11469 standards.

- Indicate the existence of toxic or harmful materials (Example 8.9)
- Use standardised materials identification systems
- Arrange codifications in easily visible places (Example 8.10)
- Avoid codifying after component production stages (Example 8.11)

8.2.5 Minimising the Overall Number of Different Incompatible Materials

Separating materials from incompatible[16] materials of the same product has its economic (due to necessary time and expenditure of the separation) and ecological (due to the impact of the input–output of such processes) costs. For this reason it would be advantageous to reduce the number of incompatible materials within one product. One of the main ways to do so is to integrate the components[17]. Equally important is to take care of the connecting elements, especially when the joined materials are compatible with each other. If the joining elements themselves are then compatible as well, the whole complex can be recycled in one and the separation process can be prevented[18].

Tables 8.2 and 8.3 show the relative compatibility of some common metals and polymers.

Guidelines for minimising the overall number of different incompatible materials:

- Integrate functions to reduce the overall number of materials and components
- *Monomaterial* strategy: only one material per product or per sub-assembly (Examples 8.13)
- Use only one material, but process in sandwich structures (Example 8.14)

Table 8.2 Relative compatibility between some metals

Metals	Elements that are incompatible for recycling	Elements that aggravate recycling conditions
Iron	Be, Hg, PCB	Cu, Zu, Sn
Aluminium	Be, Hg, PCB	Cu, Fe, Zn
Copper	Be, Hg, PCB	As, Sb, Ni, Bi, Al

From Brezet and van Hemel (1997)

[16] *Cf.* the definition in Box 8.1.

[17] *E.g.* plastic materials, because they are easily moulded and very plastic (they are not called *plastic* for nothing!) they allow to the parts that otherwise would have to remain separated to be integrated into one component.

[18] *Cf.* Chap. 9.

Table 8.3 Relative compatibility between polymers, from Severini and Coccia (1991)

	ABS	EPDM	NBR	PA	PBT	PC	PE	PEEK	PES	PET	PMMA	POM	PP	PS	PVC	SAN	SBR
ABS	ABS	5		4	4	3	5	6	6	5	1		5	5	2	1	3
EPDM		EPDM	5	5	5	5	4	6	6	5	4			5	5	5	5
NBR	5	5	NBR	4									4	4	4	4	
PA	4	5	4	PA	5	5	5	6	6	4	5		5	5	6	5	4
PBT	4	5		5	PBT	2	5	6	6	3	5	4	5	5	6	5	5
PC	3	5		5	2	PC	5	6	4	2	2		5	5	6		5
PE	5	4			5	5	PE	6	6	5	5	5	5	5	5	5	5
PEEK	6	6	6	6	6	6	6	PEEK	2	6	6	6	6	6	6	6	
PES	6	6		6	6	4	6	2	PES	5	6	6	6	6	6	6	
PET	5	5		4	3	2	5	6	5	PET	5	6	5	5	6	5	5
PMMA	1	4		5	5	2	5	6	6	5	PMMA	4	5	3	1	1	5
POM					4		5	6	6	6	4	POM			5		
PP	5		4	5	5	5	5	6	6	6	5		PP	5	5	5	5
PS	5	5	4	5	5	5	5	6	6	5	3		5	PS	5	5	
PVC	2	5	4	6	6	6	5	6	6	6	1	5	5	5	PVC	3	4
SAN	1	5	4	5	5		5	6	6	5	1		5	5	3	SAN	4
SBR	3	5		4	5	5	5			5	5		5		4	4	SBR

1 = full compatibility; 2–5 = incompatible; 6 = incompatible working temperatures

- Use compatible materials (that can be recycled together) within the product or sub-assembly (Example 8.15)
- For joining, use the same or compatible materials as in the components (to be joined)

Examples

8.12 Hitachi has redesigned its television sets to reduce the duration of disassembly, integrating several functions and reducing the overall number of components and different materials.

Example 8.12 The architecture of a Hitachi TV set before and after re-designing

8.13.1 Celle office armchair, designed by Jerome Caruso, is made on the basis of technology called "cellular suspension": a flexible polymer is moulded into a cellular structure and guarantees differentiated flexural resistance for the seat and seatback. Thus, the whole component is made of the same material and can be disassembled in 5 min.

Example 8.13.1 Office chair by Herman Miller

8.13.2 Halbert Heijn has replaced the aluminium spout of its food containers with a cardboard one; now the whole container consists of one material.

Example 8.13.2 Food packaging by Halber Heijn, before and after redesigning

8.14 GEP Plastic has presented a model of refrigerator in which the walls and door are entirely made of polystyrene. The exterior walls are made with injection moulding; meanwhile, the interior parts (the isolation) are made of foam; thus, the parts have different characteristics, but are made on the same polymer basis. In this case, it is possible to recycle the whole set of walls without previous separation.

8.2.6 Facilitating Cleaning

Normally, materials are cleaned to eliminate contaminants that remain from contact with other substances or from surface treatments[19]. In many cases it is possible to facilitate or even avoid these operations.

Guidelines for facilitating cleaning:

- Avoid unnecessary coating procedures (Example 8.16)
- Avoid irremovable coating materials
- Facilitate removal of coating materials
- Use coating procedures in compliance with coated materials (Example 8.17)
- Avoid adhesives or choose in compliance with materials to be recycled
- Prefer dyeing internal polymers, rather than surface painting
- Avoid using additional materials for marking or codification
- Mark and codify materials during moulding (Example 8.18)
- Codify polymers with a laser

[19] Organic compound additives, for example, are rarely recyclable.

Examples

8.16 Many telecommunication appliances need electromagnetic shields, a metal layer as an internal barrier. In these cases it is preferable to replace galvanic treatments of plastic coating by affixing easily removable laminates.

8.17 Labels of PVC bottles are often printed on paper and glued to the bottle. They could be replaced with PVC foil and be attached mechanically.

8.18 Polymers can easily be codified with cut or relief moulding.

Example 8.18 Moulded codification on a polymer

8.2.7 *Facilitating Composting*

Products that undergo this kind of treatment are characterised by a high percentage of decaying matter, fit for composting. Here, in general, apply the recommendations similar to those made for recycling: facilitate the collection, transportation and separation of incompatible materials. The incompatible materials here are the ones that compromise the development of compost, that is, non-organic and non-biodegradable materials.

Guidelines for facilitating composting:

- Select materials that degrade in the expected end-of-life environment
- Avoid combining non-degradable materials with products that are going to be composted (Example 8.19)
- Facilitate the separation of non-degradable materials

Example

8.19.1 Throw-away sanitary pads are mainly made of compostable materials; thus, non-biodegradable polymers should be not used for the liquid-proof parts.

8.19.2 Rubbish bags for humid waste should be made of biodegradable foil, then it would no longer be necessary to separate the waste from the bags and the whole operation would be more economical.

8.2.8 Facilitating Combustion

As said before, combustion is normally considered to be part of the cascaded approach. Thus, it is reasonable to incinerate materials after they have been recycled more than once (into components with as low requirements as possible), when their characteristics no longer satisfy any application whatsoever. Only at this point would recovery of their energy content be called for through combustion.

In some cases combustion is an efficacious way to neutralise dangerous materials that cannot be recycled[20].

Here, one again, apply recommendations similar to those made for recycling: facilitate the collection, transportation and decontamination of materials. The main problems that arise with incineration come from toxic smoke emitted by some materials and additives.

Different materials have different heat values, meaning that some materials facilitate combustion (typically the plastics, wood, paper and carton) and some hinder it – need extra energy for burning.

Guidelines for facilitating combustion:

- Select high energy materials for products that are going to be incinerated (Example 8.20)
- Avoid materials that emit dangerous substances during incineration (Example 8.21)
- Avoid additives that emit dangerous substances during incineration (Example 8.22)
- Facilitate the separation of materials that might compromise the efficiency of combustion (with a low energy value)

Examples

8.20 Glass, metals, concrete and ceramic materials all hinder incineration.

8.21.1 PVC is not considered suitable material for products that undergo combustion. Because this organic compound contains chlorine, it emanates dioxins and furan during combustion. However, better incinerators exist nowadays that are able to combust PVC too, but only these incinerators are able to neutralise these emissions.

[20] *E.g.* hospital waste.

8.21.2 Do not use chlorine-treated paper, whose combustion emits dioxins. Alternative whiteners for paper are industrial ozone and oxygen.

8.21.3 Do not use hoops coated with zinc or tin, because their combustion emits barely filterable emissions.

8.22 Do not use any paints, additives, fire-retardants or colouring pigments for plastics that contain heavy metals.

Chapter 9
Facilitating Disassembly

9.1 Introduction

Design for Disassembly (DFD) focuses on how to design easily disassembled products; meaning that the *parts* and *materials* can be easily and economically separated.

The possibility of easy separation of the *parts* facilitates the maintenance, repairs, updating and re-manufacturing of the products.

Meanwhile, the possibility of easy separation of the *materials* facilitates, on the one hand, their recycling (if they happen to be incompatible with each other) and, on the other, their neutralisation (in case they happen to be toxic or dangerous).

Therefore, the environmental arguments behind adaptation of *design for disassembly* are the *extension of product lifespan* (the maintenance, repairs, updating and re-manufacturing), the *extension of the material lifespan* (recycling, composting and combustion) and neutralisation of the toxic and harmful substances.

Design for disassembly is treated in its own chapter not only because it is, as we just saw, practical for many strategies of reducing environmental impact, but also because here the designer can play a substantial part.

In order to define *design for disassembly* efficiently, it is important to understand the typologies of the disassembly as well as the economic and environmental arguments and priorities that could be useful for its accommodation (Box 9.1).

Box 9.1 Disassembly

Here, the typologies of the disassembly as well as the economic and environmental arguments and priorities that could be useful for its accommodation are described. Due to the similarities with *design for assembly* (DFA) it is useful to start with this discipline, which has already been consolidated in many design practices for facilitating assembly.

Disassembly and Assembly

Easy disassembly also belongs to the guidelines for *design for assembly* that verify the efficiency of a well assembled product. Thus, we have already a foundation of design practice that, considering the similarities and symmetries (assembly–disassembly), could be useful for our discipline. Nevertheless, disassembly and assembly cannot be conjectured to be equal, one is different from the other; in fact, many important characters vary a great degree (Simon *et al.*, 1992).

Entropy

Disassembly – dissection of a product – starts with a single product and continues towards an increased state of disorder; it is an entropic process, meaning that, for example, in the case of material recycling, the containers and conveyors (more or less automated) are not needed for single components, but only for separating different materials.

Damaged Parts

The use of products and the disassembly processes can cause deformation and damage to the product and its components: the wear and tear, deterioration (*e. g.* rust), accidental damage and breaking up. The damages and degradation can be to some extent tolerable for the purposes of disassembly. In the case of repairs, the disassembling systems must be non-destructive, but if disassembly is meant to recover the materials, it is generally irrelevant whether the component gets damaged or not.

Disassembling Structures

The input resource of every disassembling process is the mix of discarded products (Fig. 9.1). More precisely, the disassembling structures can operate on two kinds of input.

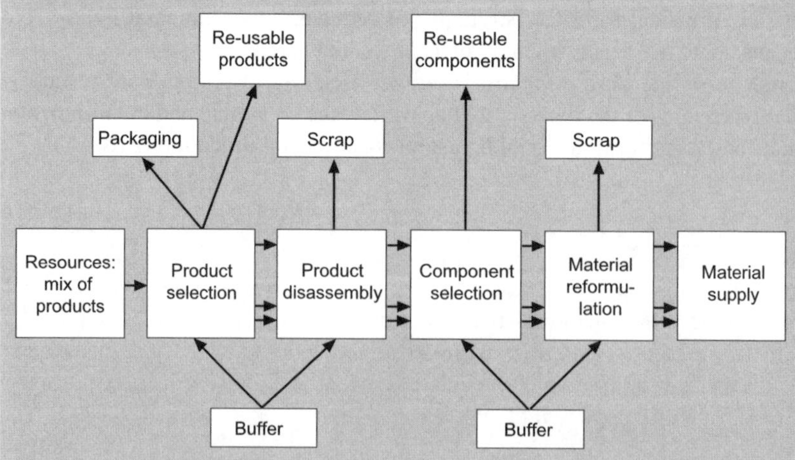

Fig. 9.1 Components and stages of disassembling structures (Simon and Dowie, 1992)

Single Products

The disassembly processes are well-defined and almost invariable. This usually concerns products that return directly to their producers.

Assorted Products

Managing and processing the product flow needs considerable flexibility. This usually concerns a recycling company that recovers products manufactured by various producers.

Due to the product assortment and their complexity, every single structure could be configured as a cellular or linear system or as a combination of the two (Fig. 9.2).

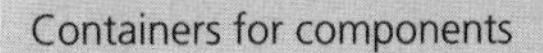

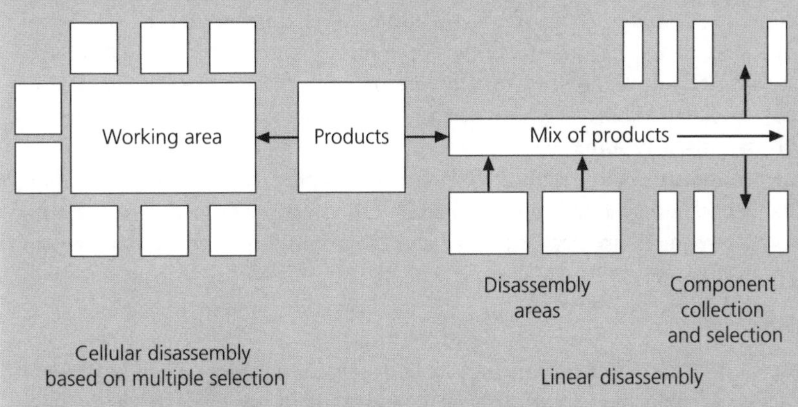

Fig. 9.2 Configuration of disassembling structures

Cellular Disassembly

The selection and separation are carried out in one place by one operator (human or robot). This kind of system has a limited number of separate containers used by the operator for the selection. The ergonomic principles indicate that for optimal selection the number of containers should not exceed ten.

Linear Disassembly

Different working stations can handle varying product flow; in this case there is no limited operational depth, since any number of containers can be arranged along the line for collection.

Parallelism

To expedite and thus make the disassembly processes more efficacious it is important, especially in the case of complex products, to have the possibility to handle simultaneously, that is, in a parallel manner, as many components as possible. Two kinds of parallelism are present: temporal and spatial.

Spatial Parallelism

Various parts and sub-assemblies can be dismantled separately, allowing various operators to work on the same product; all this increases the amount of products that can be disassembled at the same time.

Temporal Parallelism

Once separated, the sub-assemblies can be sent to another work station or along another route; this can expedite the treatments of small products and give access for specialised working cells on certain sub-assemblies, especially if their structure is sufficiently modular.

Automated Disassembly

The view inside many disassembling stations often reminds one of something from the past, especially compared with the highly automated assembling centres. Today the disassembling operations are still almost always carried out manually. Automated disassembly systems appear only in cases when no operational flexibility is required.

Still, automated disassembly is an interesting and probable option for more efficient separation systems. But compared with assembly processes automating disassembly requires greater flexibility and adaptability, for reasons that we are going to discuss now.

Flexibility

The useful lifespan of most of the products is generally greater (usually between 5 and 15 years) than the market entrance of new products, meaning that at the same time a wide variety of products, old and new, have to be processed. For this reason, the automated disassembly systems have to be sufficiently flexible.

Adaptability

The configuration of a disassembled product can depend on its age and utilisation methods, and therefore, for example, on its state of corrosion or damage. For this reason, the automated disassembling system has to be sufficiently adaptable to different stages of wear and degradation of the product.

Corresponding to what has been said above, it could be useful to outline a possible configuration of a flexible disassembly unit.

Automated and Flexible Disassembly Unit

Such a unit should be made of the following components:

- An industrial robot handling the components and disassembling machinery; typically robots with six degrees of freedom are required, together with computerised numeric control systems.
- A control computer for the disassembling process and communication, through a *computer-integrated environment* or a human operator.

- Machinery for flexible disassembly: dynamic automated screwdrivers and pick-and-place systems.
- Cutting machinery, *e. g.* laser cutters.
- Flexible features for loading and picking products.
- Tactile sensors and visual identification, laser control and for allowing the robot to react to possible alterations of the product's state (corrosion, damages *etc.*).
- Transporters for pallets and containers of disassembled parts to move them from the working cell to the cleaning, selection, repair, re-assembly, or re-manufacturing areas.

Environmental Priorities and Economic Costs of Disassembling Processes

Design for disassembly can be essential for the extension of a product's lifespan as well as for the extension of the material lifespan or for their neutralisation. Facilitating disassembly can effectuate a decrease in the costs of maintenance, repairs, updating, re-manufacturing and re-use, as well as recycling, composting and incineration.

Due to the particular importance of recycling within the product's end-of-life options we shall now explain in detail the advantages of strategies that facilitate the separation of materials.

Disassembly and Recycling

If recycling is the objective of disassembly, then the economic equation that defines the interest and profitability horizon, depends on the price of primary raw materials and costs of disposal. This dependence manifests clearly in the political-legislative and price history: the recycling processes play an increasing and inevitable role. First, the *internalisation* of disposal costs will take place so that the producer first and consumer second will directly pay the disposal costs.

Economic efficiency will be pursued by minimising the operating time (costs) and by valorising the retrievable materials.

Two purposes that could be controversial, in fact, allocating more time and energy on separation would allow cleaner materials, and therefore usually more valuable, to be obtained; at the same time, the higher costs effectuated by greater time allocation in the separation processes have to be faced. Optimisation of the time and costs of disassembly and at the same time improving material quality will be one of the outstanding issues. Generally, the materials preserve their original characteristics according to their purity, and that would determine their market price. *Design for disassembly* can offer appealing opportunities for this cause.

Recycling and Disassembly Depth

A product can be disassembled down to the separation of every single component. The degree of components' separation is called the disassembly depth, a useful variable for the analysis of recycling costs.

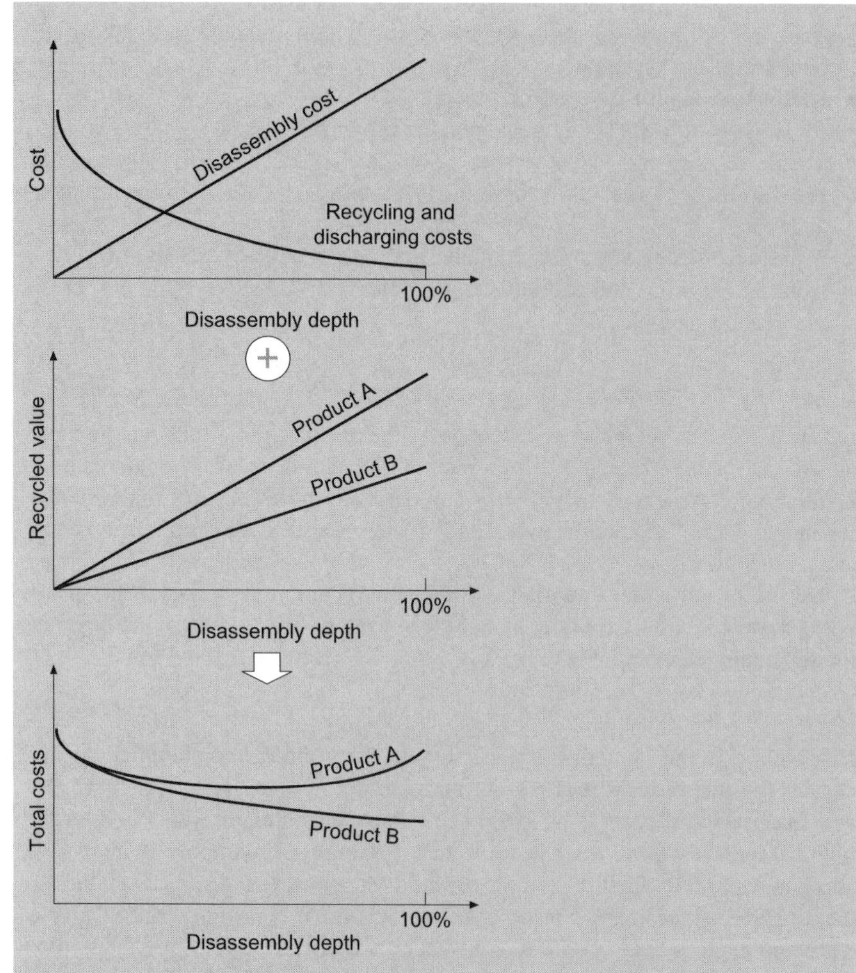

Fig. 9.3 Disassembly and recycling cost curve (Kahmeyer and Warneke, 1993)

The deeper the disassembly, the longer it will take and the higher the costs; however, at the same time the materials become more accessible, each with its own costs depending on the weight and degree of purity. Here, it is possible to define the function of disassembly depth, the final cost curve, and identify its optimal points. The actual depth to be reached with disassembly can be identified (Fig. 9.3).

Disassembly and Crushing

Adjacent to disassembly, the working theory of selecting certain materials for crushing and the following operations of separating the crushed materials should not be neglected. The main methods of automated separation are magnetic, induction and flotation, used, for example, for separating metal components from a car carcass. Obviously, this method is not valid for extending product lifespan, but only for final recycling.

There generally exist several different methods of separating the materials for recycling; these are defined by combining to different degrees the following options:

- Disassembly alone
- Crushing the entire product and separating the materials

Compared with crushing and separation, the disassembly alone can provide a better or equal quality of materials; the physical and structural qualities, thus also the economic value, of materials obtained through disassembly are better or at least equal to those obtained through crushing and separation. The comparison of costs tends to favour crushing, especially in the case of complex products.

As has been said, there are many hybrid methods that follow the disassembly and separation to different degrees.

In some cases it is useful to apply the crushing at the collection site, for example, grinding down the polymer components of some products can reduce remarkably the recyclable material volumes, thus facilitating the transportation to the re-processing facilities. The same applies to glass bottles: is useful to crush them the moment they enter – the bottle bank or the truck platform.

Methods and technologies of the disassembly and crushing are both in promising development, leaving the impression that both methods will be and will have to be followed, and the efficiency will be found out individually, case after case.

9.2 Guidelines

Depending on the product, its components and materials, the following guidelines for facilitating disassembly could be given:

- Reduce and facilitate operations of disassembly and separation
- Engage reversible joining systems
- Engage permanent joining systems that can be easily opened
- Co-design special technologies and features for crushing separation

The following guidelines apply, if after separation the materials are partly or completely crushed.

- Use materials that are easily separable after being crushed
- Use additional parts that are easily separable after the crushing of materials

9.2.1 Reducing and Facilitating Operations of Disassembly and Separation

The indications for minimising and facilitating disassembling operations depend on the general architecture, the shape of components and parts and finally on the accessibility to the junctions.

When we examine the architecture and the structure of the connections, it is important to understand how the components could be replaced without disassembling other parts; or how to carry out parallel disassembly. Modularising could be a very efficacious strategy, especially for product life extension and particularly for the upgrading, repair and re-use of the components.

The linearity of the dismantling operations is essential for automated disassembly.

It is generally recommendable to minimise the overall number and different kinds of joints, as this helps to reduce disassembly time. Naturally, it has to be compatible with the structural requirements of the product or component under examination.

Finally, it is important to provide proper information on the disassembly procedures, especially in the case of open-loop recycling. As in this case, the disassembly operations are carried out by entities that were not the manufacturers and could not possibly know the exact configuration of the product.

Guidelines for reducing and facilitating operations of disassembly and separation:

Overall architecture:

- Prioritise the disassembly of toxic and dangerous components or materials (Example 9.1)
- Prioritise the disassembly of components or materials with higher economic value (Example 9.2)
- Prioritise the disassembly of more easily damageable components
- Engage modular structures
- Divide the product into easily separable and manipulatable sub-assemblies (Example 9.3)
- Minimise the overall dimensions of the product
- Minimise hierarchically dependent connections among components
- Minimise different directions in the disassembly route of components and materials
- Increase the linearity of the disassembly route
- Engage a sandwich system of disassembly with central joining elements

Shape of components and parts:

- Avoid difficult-to-handle components
- Avoid asymmetrical components, unless required

- Design leaning surfaces and grabbing features in compliance with standards (Example 9.4)
- Arrange leaning surfaces around the product's centre of gravity
- Design for an easy centring on the component base

Shape and accessibility of joints:

- Avoid joining systems that require simultaneous interventions for opening
- Minimise the overall number of fasteners (Example 9.5)
- Minimise the overall number of different fastener types (that demand different tools) (Example 9.6)
- Avoid difficult-to-handle fasteners
- Design accessible and recognisable entrances for dismantling
- Design accessible and controllable dismantling points

Examples

9.1 The cooling circuits of refrigerators that contain substances hazardous to humans or nature should be positioned so that they can be accessed directly from outside.

9.2 Printed circuit boards contain recyclable high value elements like silver, gold and chromium; toxic elements like beryllium and mercury; and other elements that could compromise the recycling processes like organic electrolytic solutions inside the condensers (which corrode metals during recycling). Therefore, the separation of toxic and hazardous materials has to be facilitated, as well as the separation of high value elements.

9.3 *Cab* chairs, designed by Mario Bellini in 1977, have an easily separable structure of a steel support and a leather cover.

Example 9.3 Cab chair by Cassina, designed by Mario Bellini

9.4 Power Macintosh G5 is a computer workstation that has been designed for easy upgrading. Easy accessibility and visible internal parts facilitate the replacing of components, such as video cards and RAM, by the user himself, without the need to call for technical assistance, which may turn out to be too expensive and could be avoided.

Example 9.4 Apple PowerMac G5 workstation, equipped with easily replaceable components

9.5 *Dry* chair, designed by Massimo Morozzi, has a structure of 20 wooden components fastened by one screw; no other fastening materials (glue, nails) are used.

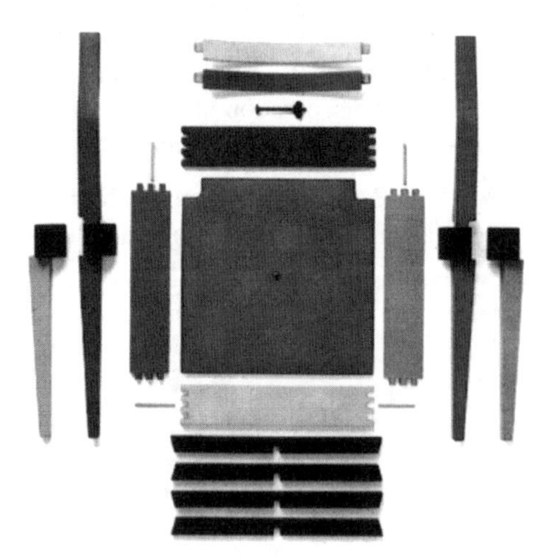

Example 9.5 Dry chair, designed by Massimo Morozzi

9.6 Use only screws with similarly shaped heads.

9.2.2 Engaging Reversible Joining Systems

From the disassembly point of view, the joints are classified as *reversible* or *permanent*. Unlike permanent ones, reversible joints can be removed and re-used without damaging the joint itself.

It is easy to realise that reversible joints are more convenient for disassembling. Especially if it is disassembly for extending a product's lifespan, then damaging the components would compromise their re-use. In respect of extending the material lifespan, this issue is not so strict; in fact, if of recycling is more important to acquire the materials as *clean* as possible and unbroken, the components are not end in themselves. Then the joining system can be permanent, as long as the materials are easily separable.

Let us take a look at some indications with regard to different reversible joining systems:

Snap-fit assembly: two way[1] snap-fits are probably ideal joints for DFD strategies, as they allow easy, fast and repeated separation of components. Besides, they have integrated with components and therefore do not increase the amount of components or compromised materials (Fig. 9.4).

Screws and bolts[2]: we should give some clarification about using screws. Compared with snap-fit assembly, a screw obviously increases the number of components and different materials[3]. But actually, it is a good reversible fastening system and appropriate for frequent disassembling/re-assembling, which is especially recommendable in cases of components that are frequently repaired and replaced.

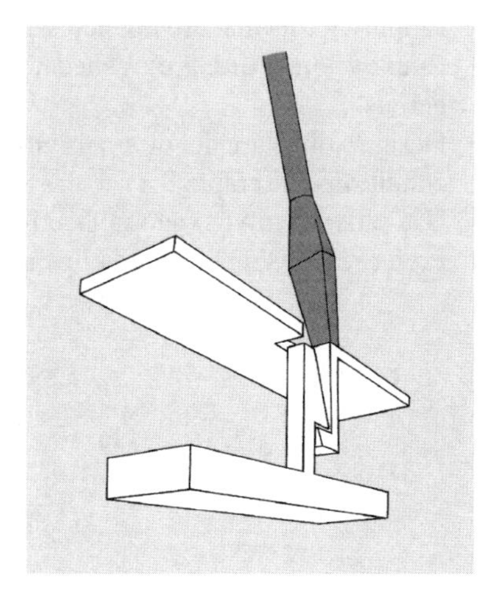

Fig. 9.4 Two-way snap-fit that is easily opened with a screwdriver

[1] Two-way snap-fit can be opened from the outside with a tool that lifts the snap's hook. For this, one of the two components must have a fissure to allow the tool (usually a screwdriver) to be pushed in and then to lift the components apart.

[2] Here, bolts refer to the set of screw, nut and washer.

[3] This problem might not even arise if the component consSolls of the same material as the screw.

Fig. 9.5 A screw with an insert

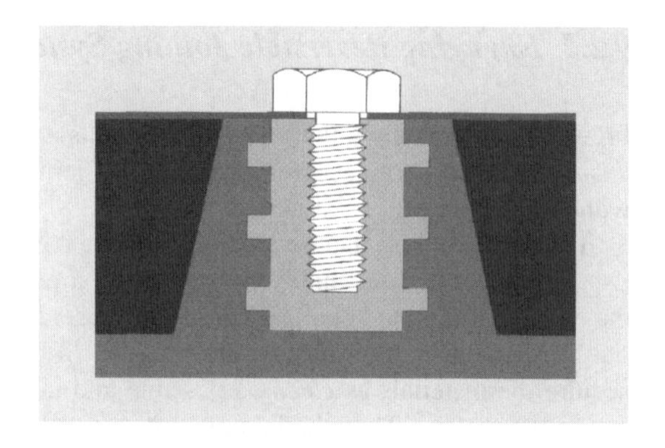

Metal screws are contra-indicated for polymer recycling and can create several problems (Fig. 9.5). At last, some screw heads (*e. g.* hexagonal) are removed more easily than others and do not any require previous cleaning operations.

Inserts (for screws): an insert is a supplementary unit that is hardly removable and usually compromising; it should be employed only if no other solutions are available. It is recommended to facilitate the break-off of the insert before the recycling process (Fig. 9.6).

Guidelines for engaging reversible joining systems

- Employ a two-way snap-fit
- Employ joints that are opened with common tools
- Employ joints that are opened with special tools, when opening could be dangerous
- Design joints made of materials that become reversible only in determined conditions (Example 9.7)
- Use screws with hexagonal heads
- Prefer removable nuts and clips to self-tapping screws

Fig. 9.6 Pre-disposed break-off areas

- Use screws made of materials compatible with joint components, to avoid their separation before recycling (Example 9.8)
- Use self-tapping screws for polymers to avoid using metallic inserts

Examples

9.7 Brunel University, UK, has developed special screws made of polyurethane-Shape Memory Polymer (SMP). Its elasticity is very modular and is subject to differentiated temperature (Tg); consequently, these screws can be easily disassembled when exposed to critical temperatures.

Example 9.7 Fastening screws made of polyurethane SMP

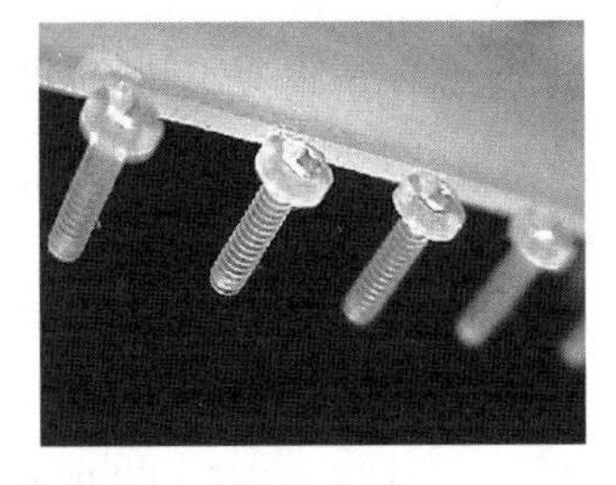

9.8 Metal screws in plastic materials pose serious recycling problems. If normal depreciation (which effectuates frequent disassembly and re-assembly) does not interfere it is recommended to use thermoplastic screws made of polymers that are compatible with the components.

9.2.3 Engaging Permanent Joining Systems that Can Be Easily Opened

It has been said that reversible systems are generally better; if the permanent systems are more convenient, then according to the chosen system the following indications should be taken into account:

Rivets: leaves behind contamination marks, when the components are incompatible with each other.

Hot riveting: preferable to nailing, because it is less intrusive, being integrated with one of the connecting components. Providing necessary fastening force requires comparatively little material, just using raw force is sufficient for fast separation.

Pressurising systems: can pose serious problems for disassembly; separable with physical force.

Welding; separable with physical force, though it often occurs that welded components are compatible and in the case of recycling there will be no need for separation.

Solvent welding (of polymers)[4]: preferable to adhesive bonding as it does not bring along additional materials.

Adhesive bonding: induces contaminants and waste, especially when applied to polymers; separable only with physical force.

Guidelines for engaging easily disintegrated permanent joining systems:

- Avoid rivets in incompatible materials
- Avoid staples in incompatible materials
- Avoid additional materials while welding
- Weld with compatible materials
- Prefer ultrasonic and vibration welding with polymers
- Avoid gluing with adhesives
- Employ easily removable adhesives (Example 9.9)

Example

9.9 Use water-soluble adhesives when labelling bottles.

9.2.4 Co-designing Special Technologies and Features for Crushing Separation

Crushing disassembly technologies are not generally adaptable for *extending product lifespan*, but they give an efficient result when fast separation of materials is required, or, as in the above-mentioned case, when removing non-compatible screw inserts. Obviously, such break-off points should not cause unexpected fractures during use.

Guidelines for co-designing special technologies and features for crushing separation:

- Design thin areas to enable the break-off of incompatible inserts with pressurised demolition
- Co-design cutting or breaking paths with appropriate separation technologies for separating incompatible materials (Example 9.10)
- Equip the product with a device for separating incompatible materials (Example 9.11)

[4] Often possible only with amorphous thermoplastic polymers and hard to apply to crystalline or semi-crystalline polymers.

- Employ joining elements that can be chemically or physically destroyed (Examples 9.12)
- Make the breaking points easily accessible and recognisable
- Provide the products with information for the user about the characteristics of crushing separation

Examples

9.10 Marco Capellini carried out research for Nordica on water-pressure separation of the ski-boot materials. The separation is conducted by applying water-jet technology to the joints.

Example 9.10 The route of a water-jet for material separation

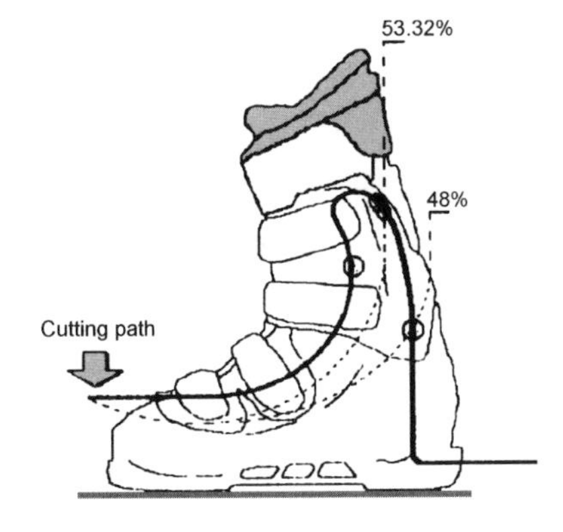

9.11 Stanford University has developed a TV set design in which during production a nichrome wire is inserted into the CRT in order to facilitate separating processes and decrease recycling costs.

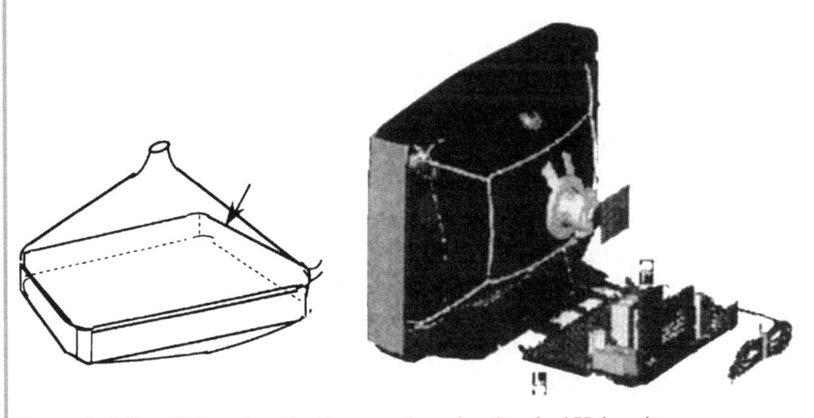

Example 9.11 CRT equipped with separation wire, Stanford University

9.12.1 Drilling can be used for removing permanent fasteners and snap-fits and releasing separated parts.

9.12.2 FIAT has designed a special cavity in the fuel tank of its cars to facilitate removing the suspenders that attach it to the bodywork.

Example 9.12.2 Fuel tank feature that facilitates its removal, FIAT

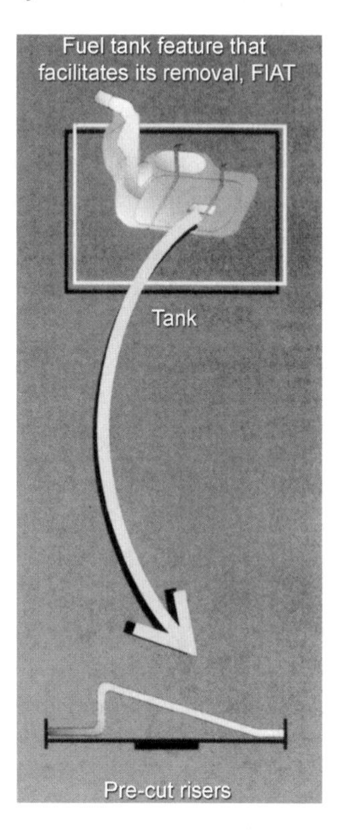

9.2.5 Using Materials that Are Easily Separable After Being Crushed

This guideline is valid from both an economic and a technological perspective. Proper technologies for separating ground materials[5] has to be known very well, as for example from this perspective we can combine (when diverse materials are necessary) metals and plastic, ferrous metals with non-ferrous metals or two polymers with remarkably different densities.

---

[5] *Cf.* Box 9.1.

9.2.6 Using Additional Parts that Are Easily Separable After the Crushing of Materials

The discussion we had on materials also applies to different kinds of inserts. It is necessary to keep the incompatible inserts easily separable with recycling technologies (Example 9.13).

Example

9.13 Thanks to the separation systems available, the ferromagnetic metals (iron, steel, nickel) are easily separable compared with metals separable with induction methods (*e. g.* aluminium).

Chapter 10
System Design for Eco-efficiency

10.1 Economic Restrictions in Traditional Supply and Demand System

Would it be comforting to claim that an approach to product Life Cycle Design (that is, reducing both the consumption and emissions/waste) can also have economically sound results? Here, in fact, we would not have to worry, but unfortunately the degradation and increasing risks to our ecosystems prove otherwise.

It is a fact that resources and energy have their economic costs as well as their environmental ones; therefore, reducing their consumption is a source of saving, just as well as it is a fact that emission and waste treatments incur steadily growing costs.

However, these facts alone are not enough to *spontaneously* [1] motivate the demand satisfaction system to considerably reduce the environmental impact of our production–consumption system. More precisely put, these calculations do not apply equally at different stages of the product life cycle. The demand–supply system within a life cycle is complex and is characterised by different stakeholders, but single interests do not necessarily coincide with the interests of a likely sustainable society.

To shed light on what was just said, it could be useful to examine some aspects of the mutual relationships between who manufactures the goods, the materials, who distributes the products, who uses them, who disposes of them and, on the other side, who would benefit from the reduction of consumption and emissions during different stages of the life cycle.

We are trying to explain these economic conditions, using the same life cycle scheme, but replacing different stages with different stakeholders (Fig. 10.1). With a careful inspection of the traditional supply market, we can observe that in general terms the economic interests motivate the stakeholders to undertake resource optimisations on the *sub-system* level. For example, anyone who manufactures the

[1] That is, without legislative interventions that would favour some processes and hinder others, as in the case of *eco-taxes* and different *eco-incentives*.

materials (pre-production) would be inclined to reduce the material consumption rate per unit (inner arrow in the figure), but would be later motivated to sell (and pre-produce) as many materials as possible.

For the same economic reasons, anyone who manufactures the product (the washing machine in our example, inner circle in the figure), normally engages innovations for reducing the material costs (consumption) as well as emissions during the production stage (eventually also during the distribution) per unit produced, and then the dominant logic on the market incites the producer to sell (produce) as much as possible.

Reducing the factor costs and emissions during use are not necessarily among the business goals. Meanwhile, the characteristics of potential efficiency of the utilisation (reducing consumption and emissions) and the environmental valorisation of discards are both determined during the development stage.

And later, during disposal, the costs are not linked to the product (internalised), but are paid for by the society (usually with tax money).

In summary, it can be pointed out that there are direct interests behind reducing the consumption during *the transformation processes within a phase* (operated by one economic *actor*), but not during the *phase transition* (passing from one economic *actor* to another).

Now, in the traditional supply model, that is, when the supply model is focused solely on sales of products or semi-finished products, the problems of adapting an LCD approach, that is, minimising the resource consumption and system level impact, derive from the *low-level interaction between product system stakeholders* (or rather the *low-level interaction between satisfaction system stakeholders*).

This means that the biggest problems do not appear in the *transformation processes* within one given phase, when they are within the competence of a single stakeholder.

More problems might arise in the so-called *phase transaction*, during the sale or disposal of semi-finished products and products from one stakeholder to another, where economic interests appear that can motivate to increase resource consumption. For example, a producer of washing machines has an interest in reducing material and energy consumption during the production phase. On the contrary, he has no direct economic interest either in limiting consumption during use, or in reducing divestment impact and valorising the resulting waste. Sometimes the producer is even interested in selling products with a short lifespan, with the only aim being accelerating replacement. We can generalise this by saying that in a traditional supply system, there are direct incentives to reduce the consumption during the *transformation processes* within a phase, but no interests in decreasing or even interests in increasing the consumption during the *phase transition*.

Similar problems can be found observing the product service systems that satisfy a certain demand; the so-called *combinations of life cycles* in a traditional supply system in which stakeholders do not have any direct interest in reducing resource consumption.

In summary, the fragmentation of stakeholders in the various phases of a product's life cycle (within the traditional economic framework of industrialised countries)

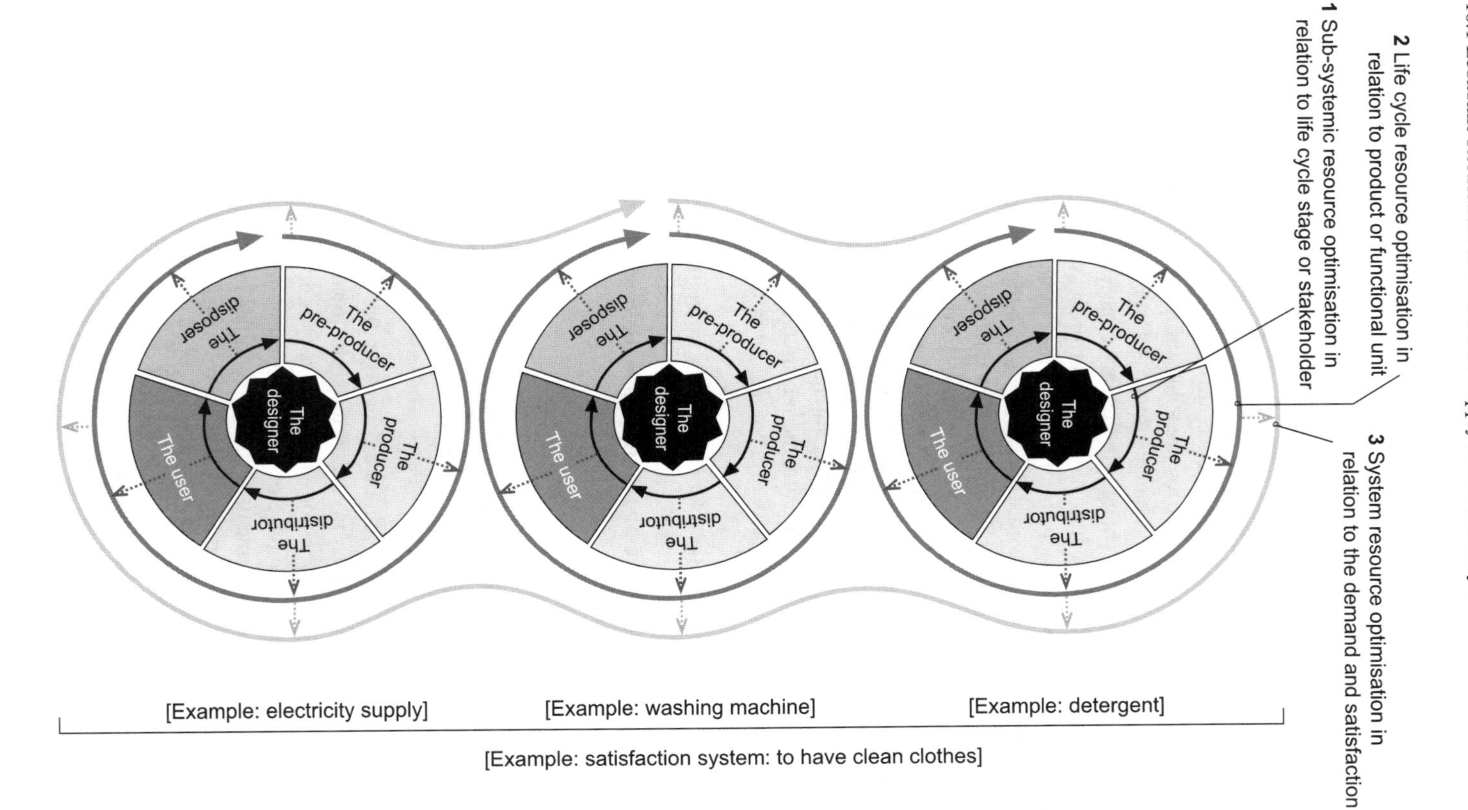

Fig. 10.1 Different levels of resource optimisation in conjunction with the economic interests of the life cycle stakeholders

means that the eco-efficiency of the life cycle system does not usually coincide with the economic interests of the individual constituent stakeholders.

This observation is true both in the case of the producers of the product service system that participate in a certain demand satisfaction, and in the case of the producers who do not, but who are connected with the cascaded use of emissions and discards within productive processes.

As we have underlined more than once, this framework refers to the traditional economies of industrialised countries, focusing on the sales of physical products. The urgency and importance of changes towards sustainability, however, force us to argue over possible radical changes in the very concept of the offer itself.

Here we ask if, within the innovative market processes, alternatives exist that could shift the production–consumption activities towards such a production–consumption system, which would drastically and altogether reduce the environmental impact at all stages of the life cycle that relate to the supply.

Returning to the scheme of life cycle *actors*, it would be important to find such modalities of the relations and interactions (already present or still in development), which would motivate, also through direct economic incentives, the optimisation of the life cycle, or even change the product's functional unit (middle circular arrow in Fig. 10.1). Or even better, interactions between different economic *actors*, which could lead to optimising the resource consumption within the entire product–consumption system behind the *satisfaction* of certain demand (external arrow in Fig. 10.1).

When, as in our scheme, the "satisfaction" is in having clean clothes, then we are interested not only in the washing machine, but also in detergent, water and electricity services, and above all, in the economic *actors* who supply them and interact with each other.

10.2 System Innovation for New Interactions Between Socio-economic Actors

If what is said is true, which characteristics should the supply (semi-products, products and services) have, in order to shift towards greater system eco-efficiency, that is, towards situations in which economic and competitive reasons would compel businesses to reduce resource consumption at a systemic level? One has to look for innovative elements in the stakeholder interactions and *configuration* between different stages and life cycles of the satisfaction system.

Here, a vertical stakeholder integration could help, a single *actor* who is or could become responsible for more than one stage within the life cycle, *e. g.* a washing machine producer who takes care of their disposal. Or a horizontal integration, a stakeholder who is or becomes responsible for more than one life-cycle within a satisfaction system, *e. g.* a washing machine producer who also produces washing powder and later takes care of their disposal.

Vertical and horizontal integration has its own limits due to monopolistic risks that can effectuate other inefficiencies (and social injustice) enabled by the absence of competition.

However, the extension of control is not the only way to introduce *cross-phase* and *cross-cyclic interactions*, various forms of partnerships and services could also become helpful, that extend the liaisons between different economic actors (including the user[2]).

Thus, the relationship between the supplier and buyer would not end with the moment of purchase (semi-product or product), but perpetuates over time. Also, these partnerships could be vertical or horizontal.

10.3 The Supply Model of the Product Service System

When speaking of supply models that look to higher system eco-efficiency, as in our discourse, or when referring to the studies of international research centres, terms like *system innovation* or *Product Service System (PSS) innovation* are used.

Using the definition from the United Nations Environmental Program (UNEP, 2002), it can be claimed that it is "the result of *an innovation strategy, shifting the business focus from designing and selling physical products only, to selling a system of products and services which are jointly capable of fulfilling specific client demands.*"

Here, we are talking about eco-efficient system innovation effectuated by a new convergence of interest between the different stakeholders: not only product level innovation, but also innovation between different stakeholders (new forms of interactions/partnership). Dealing with innovation for an economic scenario inside the so-called "new economy", characterised by the movement from the system focused on product sales towards another in which supply is configured as an integrated mix of products and services; a supply system in which the producer remains an owner of the product for the entire life cycle.

Such a scenario has some remarkable potency for eco-efficiency. As we are going to see, within this framework the producer can also profit from minimisation of resource consumption, emissions and waste during the use and disposal and is motivated to develop durable goods that require few resources and emit little waste.

To begin with, we have to understand the main economic springs that in such a scenario would push the supply model to use Life Cycle Design strategies.

First of all, for the intrinsic factor of service economies: the producer (supplier) is interested in reducing the operating costs, including resource consumption, during the entire life cycle, including use and disposal.

This is seconded by the motivation to extend product life, in order to postpone the disposal costs and new production costs (of products that replace the futile

[2] On this subject, *cf.* Part I.

products). He would also save on production by re-using some parts of discarded products, because these are already at his disposal.

Because the supplier remains an owner of the goods, he would have to look for proper valorisation of the disposed materials, so that he could save on disposal costs as well as on purchasing new materials.

Moreover, the potential collective nature underlying services determines more intensive engagement of products and materials, that is, in general terms, contrive to subdue the overall number of products required simultaneously by communities. If a product is employed by more than one person, then fewer products would be demanded at the same time in the same place[3].

Another advantage arises from the greater professionalism and heightened level of technological progress that are economically available for product service providers; surpassing the barriers created by high, unsustainable investments with more eco-efficient technologies.

And in the end, there are cases in which economies of scale, effectuated by a high user base, allow considerabe emissions and resource consumption.

In conclusion, given a certain demand, the economy based on a product service system, which transfers the ownership of physical goods from the user to the producer, is essentially more eco-efficient than an economy based on individually produced and consumed goods. Indeed:

- More efficient resource consumption: minimisation plus selection of low impact resources
- More intensive use of products
- Direct interests to extend product lifespan
- Direct interests to extend material lifespan

It will not be all that easy: it has not been said that service providers will operate in economic and environmental unison with the users. But it should be noted that nowadays the service is less economical than personal ownership over the product, or at least it is considered so[4]. And finally, not all services are necessarily eco-efficient either.

10.4 Guidelines

The following describes some supply strategies for developing eco-efficient offers:

- Services providing added value to the product's life cycle
- Services providing "final results" for customers
- Services providing "enabling platforms for customers"

[3] Actually this issue is a bit more complex, as the lifespan, functional quality and maintenance requirements are subject to the to some extent intensive shared usage.

[4] Is it enough to consider the costs per kilometre for the car owner: the cost we do take into account is fuel consumption, but we forget the amortisation, repairs or maintenance. Try to tell someone that it is more expedient to go shopping by taxi; they would laugh in your face, ignoring the additional costs of the car.

10.4.1 Services Providing Added Value to the Product's Life Cycle

We can speak about services providing added value to the product life cycle when a company (or alliance of companies) provides additional services for the final user together with the purchase of goods or semi-products connected with its life cycle stages – production, distribution, use and/or disposal.

This kind of service can lead to new relationship between the client and provider beyond the moment of purchase. The following examples illustrate how and why the companies are economically motivated to create eco-efficient systems from this perspective (Examples 10.1–10.3).

Examples

10.1 Klüber Lubrification has started to offer additional services to lubricant sales, such as onsite identification (at the industrial sites of its customers) of the equipment efficiency and of possible reductions in emission impacts (due to the use of lubricants). For this purpose they are using a properly equipped movable chemical laboratory that is able to monitor in a working environment (industrial machines) the performance of lubricants used and their environmental impact. There has been a reduction in lubricant consumption and polluting emissions, through a supply system that has extended relations with the client.

The activities have also effectuated a reduction in lubricant sales, but the losses are covered by service fees charged by Klüber.

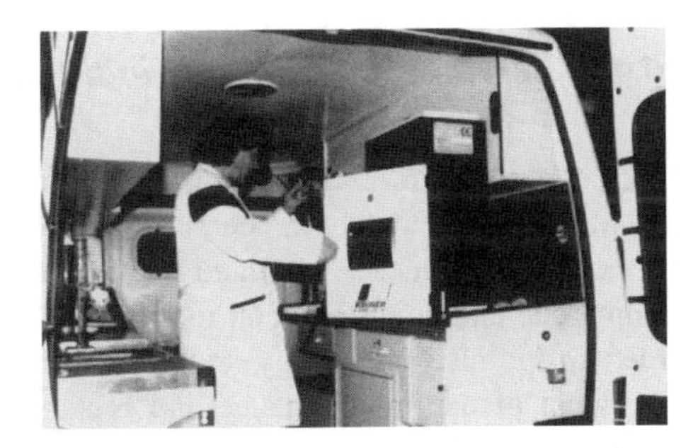

Example 10.1 Lubricant sales combined with the service of consumption optimisation, Klüber

10.2 Diddi & Gori provide a service of supplying, installing and removing textile floors for fairs and exhibitions. They have acquired easily recyclable textile flooring, made of post-consumer discards; this replaces the traditional engagement of products that were used for the short term and then dumped, thus reducing the management and environmental costs.

10.3 Allegrini is an Italian cosmetics and detergent producer that since 1998 has provided a Casa Quick service of detergent home delivery.

Casa Quick products are distributed from vans that drive from house to house according to their delivery route. According to their needs, every family can refill their reusable plastic containers (provided by Allegrini) and pay for the just right quantity of detergent. At the same time the clients can acquire information about optimising detergent consumption.

Before the introduction of this service the customers were using throw-away bottles, but the re-usable ones that are essentially a part of the service have drastically reduced the need for packaging (and reduced the material consumption).

Besides, other problems connected with recycling traditional containers that are caused by the detergent residuals are avoided, which may be in danger of contaminating the environment if discharged.

Example 10.3 Detergent home delivery with reusable containers, Allegrini

10.4.2 Services Providing "Final Results" for Customers

This supply system anticipates the transition from product sales to result sales (Examples 10.4 and 10.5). The supplier here provides a service and remains the owner of the product. As pointed out before, in this case, the company has all the commercial and environmental motivation to continually improve performance in order to optimise the energy and material consumption during use. Besides, the placement of use, maintenance and repairs under professional management has efficacious results.

Examples

10.4 Rank Xerox is a photocopier producer. In Germany it is offering package deals of photocopiers, their maintenance and repairs, as well as doing the actual copies, collecting the originals and distributing the copies. For this package it has developed a systemic approach – the Chain Management System – to deal with the re-use and recycling of equipment components.

This system includes designing, disassembly, the tests before re-use, recycling of materials and assembly. Discarded photocopiers are disassembled in a special factory, the components undergo exhaustive tests and the ones that pass are used in new photocopiers. These copiers have exactly the same characteristics as if they were completely new, since they respond to the same requirements (pass similar tests). Damaged parts, meanwhile, undergo material recycling. In this method about three-quarters of the components can be re-used, with some parts being up to 90% recyclable.

Example 10.4 Component reutilisation workshop of Rank Xerox photocopiers

10.5 AMG Palermo, Italy, is an electricity supplier. For some years now it has provided a specific service for pay-per-use water heating. The water is heated in a hybrid boiler house fed by solar and methane-derived power. In the end, the service includes: methane supply (indirectly paid for), machinery (not in the client's ownership), its transportation, installation and maintenance.

It is remarkable that the income of the provider depends greatly on the methane consumption. This motivates AMG to provide more efficient machinery and favour solar energy (which entails no direct costs).

Example 10.5 Provider of heated water, AMG Palermo

10.4.3 Services Providing "Enabling Platforms for Customers"

This approach refers to business activities that provide access to products, equipment or possibilities in general (*platforms*) that allow the customer to obtain the results demanded. It means that the customer works autonomously to engage the utilities, but does not own the products, paying only for the service provided by the products. The usual commercial applications of this kind are *leasing*, *sharing* and *renting*.

These supply methods can lead to intensified use of products, as probably more than one customer would be using the same product (Examples 10.6–10.12). This has the evident advantage of reducing the overall number of products simultaneously present in a given place[5].

Also, here the provider of the collective service is more efficient because it has greater professionalism, access to a higher level of technology and can benefit from economies of scale.

Usually, these kinds of services require certain reasonable changes in one's lifestyle and in one's perception of well-being.

Examples

10.6 Launderettes have environmental advantages compared with domestic washing machines and these are effectuated by the higher efficiency of industrial washing machines, due to their technological advances and the higher amount of laundry processed at the one time. Current disadvantages reside in the transportation of laundry (to and from the launderette) and energy consumption during mechanical drying.

Various structures could be employed for stationing communal washing machines and disengaging the consumer's active participation:

- Service centres: service networks in residential zones that offer professional performance
- Integrated into a communal room inside a building: allows advanced technologies to be used that otherwise would prove too expensive
- Laundrette combined with other activities: offer the services away from home, in places that are frequently used due to work or travel
- Laundry home delivery service: dirty laundry is picked up and returned home

---

[5] The advantage is not so much in reducing the overall amount of products produced and disposed of, but their exact presence at a given time and space. It is hard to say straight away if the actual turnover of production and disposal is going to be reduced or increased: a more intensely used product tends to become worn out sooner. But a reduction in the simultaneous presence is an advantage indeed. It is enough to think about traffic jams – the same service with fewer cars present pollutes less than an intense traffic situation.

All these services require from the customer certain changes in their lifestyle and in their perception of well-being. The domestic washing machine is held as a commodity; new services, therefore, have to appear even more attractive.

10.7 Car pooling is a service organised by people who share the same travelling routes to some extent, so they can travel together using the same car. This reduces the total mileage of cars and traffic density. Environmentally, it results in a reduction of smog and its effects.

10.8 Car sharing is a service of a shared car fleet. Having access to the car without owning it reduces remarkably the overall number of cars in circulation and parked on roadsides or sidewalks. Personal advantages are the lower costs (pay per kilometre) and a greater choice of cars available for different situations.

Example 10.8 MilanoCarSharing, Legambiente, Italy

10.9 Call-a-bike is a public urban mobility system (in Germany) that gives its members access to specially designed bikes. Once registered, the customer can pick up and leave bikes at all major junctions of the city with just a phone call. Bikes have electronic locks that can be opened with authentication codes. When a client finds a suitable bike he/she just has to call the number indicated on the lock and receive the authentication code.

Example 10.9 Call-a-bike, Germany

10.10 Virtual Station is a Brazilian company that offers its clients (independent professionals, commercial representatives, individuals or third sector entities) access to a "virtual office". They offer complete office space, equipment, services and infrastructure; the customers only pay for the time and services they require. Sharing equipment and office space allows them to be used more intensively, and reduces the overall number of products necessary for individual users.

Example 10.10 Virtual office, Brazil

10.11 A partnership between Ariston and Enel (Italy) has created a new *pay-per-use* washing machine service. It is based on a prepaid number of loads and includes: delivery of the washing machine (which remains in Ariston's ownership), electricity supply (no direct costs), maintenance, upgrades, replacements and end-of-life collection. This innovative relationship among Ariston, Enel and the end-user economically motivates highly efficient equipment to be designed that is upgradeable, easily maintained and repaired, and recyclable.

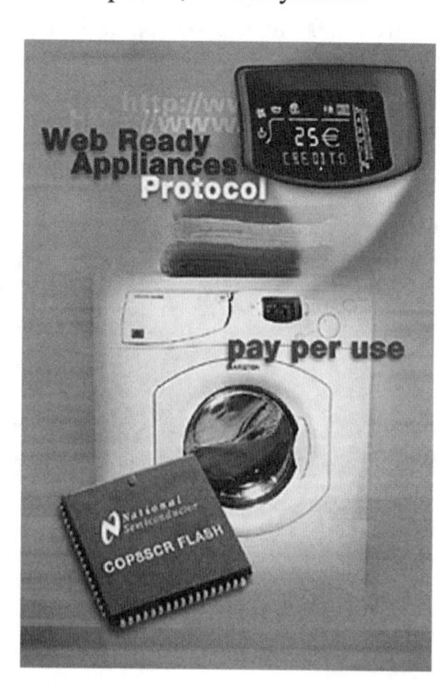

Example 10.11 Pay-per-use washing machines, Merloni and Enel, Italy

10.12 Greenstar is a company whose primary objective is to satisfy different demands in rural areas of developing countries that have no access to electricity and modern communication technologies.

The second goal is to turn Greenstar stations into income sources for local inhabitants, giving them access to e-commerce and the world markets where they can sell their art, handicrafts and local agricultural products.

Greenstar installs multi-purpose modular-built work stations that have high-speed internet connections, a small medical clinic and a lecture room.

Centres are equipped with a solar panels system that provides all energy necessary for its functioning (autonomous installations fed with renewable power). Solar energy can additionally be used for water purification via thermal distillation and pasteurisation.

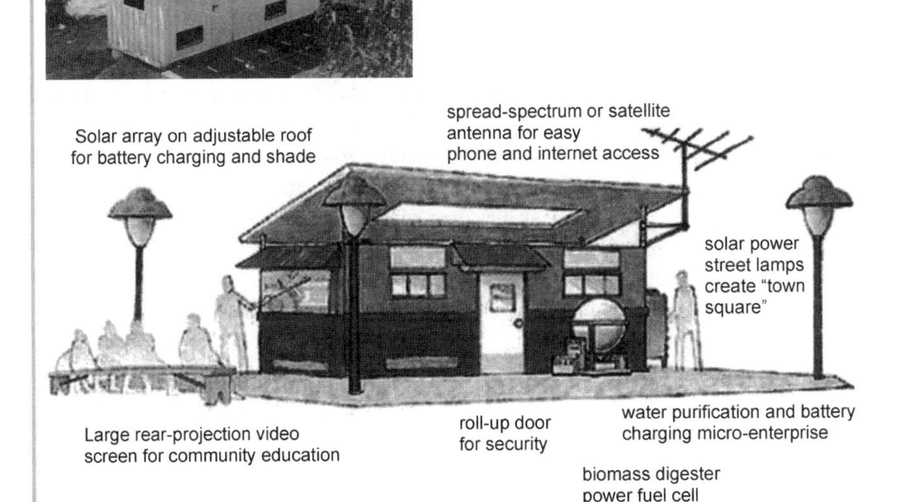

Example 10.12 Solar-powered e-commerce and community centre, Greenstar

10.5 Strategic System Design for Eco-efficiency

We have seen now which opportunities can give system innovation to the sustainable product and service development. Returning to the role of designer we should first explain, which kind of innovations we are speaking of here.

They are radical innovations, not in the strict technological sense as much as in new interactions and partnerships between different stakeholders of the satisfaction system (life cycle(s)). Innovations that can lead to convergences of economic interests between the participants are characterised by system eco-efficiency.

Because of this, introducing system innovations within design for sustainability requires new skills. First, it means that we have to learn to design integrated products and services. This brings up an issue that is relatively new in current design practice and requires designers to learn how to design the stakeholder configuration, in order to find solutions that might combine the economic and environmental interests. It is not so much the technical details of the products as the relationships between different stakeholders that have to be designed.

It has also been said that in the end we have to learn and design in a design environment together with various companies, NGOs, public institutions and, of course, users[6].

All these new abilities have to rejoin design activities with strategic design, demanding competence for participating in design and for identifying promising *interactions* between the *actors* of a determined value constellation[7].

These developments have already started to lead to a convergence of design and environmental sustainability and the *strategic design* field. The term *strategic design for environmental sustainability* has been coined, which coincides with the term *system design for environmental sustainability*. But the main point is that design for environmental sustainability has to acquire and integrate the methods and tools of *strategic design* or *system design* (and *vice versa*)[8].

---

[6] *Cf.* Vezzoli *et al.* (2006); Vezzoli (2007).

[7] *Cf.* Normann and Ramirez (1995).

[8] Analysis of this topic was in Part I.

Part III
Methods and Support Tools for Environmental Sustainability Analysis and Design

Chapter 11
Environmental Complexity and Designing Activity

11.1 Introduction

In this part we describe the supporting methods and tools for analysis and design for environmental sustainability.

Including the environmental variable in the design process makes the whole activity more complex (and interdisciplinary). New requirements (environmental ones) have to be taken into account and extended along all the stages and among all the *actors* of the life cycle. Actually, one has to have more information and assess more relationships:

- Between the production–consumption system and the environment
- Between different *actors* of the product development system
- Between the latter and the *actors* involved in the products' life cycle

Within this framework of greater designing complexity, the information technologies acquire an edge, because they have the capacity to contain, circulate, compare, elaborate and present in various forms (and interfaces) the immense amount of information needed, because of their capacity to manage the increased complexity.

The indication for information technology to be supportive in designing solutions for environmental issues should by no means be conclusive. In fact, the changes required by the transition to sustainability are systemic and demand not only technological but also social and cultural changes.

From the analysis of how environmental issues and Life Cycle Design have entered into the product development processes, two approaches arise.

On the one hand, quantitative analysis of the environmental impact of a product's life cycle has found its place. This tool has come into being due to the demand for qualitative assessment as well as from the necessity to assess and compare alternative proposals. *Life Cycle Assessment* (LCA), among others, is well established and will be analysed in the second chapter of this part.

On the other hand, we have several tools for orientating design decisions; these instruments are described in the third chapter.

11.2 Methods and Tools for Design for Environmental Sustainability

Obviously, the designer has to have from the very beginning proper information and decision support tools in order to avoid illusory decisions or strategies and to visualise and focus with certain reliability on the most important fields in which to intervene. That is, the designer has to have instruments for analysis and assessment, together with tools for decision orientation.

To understand which support tools might become useful for the designer we should take as a background a simplified scheme of different stages of a product, service or system development (Fig. 11.1).

In Fig. 11.1 we outline the development stages that define the concept, develop an executive project and end with engineering.

This scheme describes product design, but is equally valid in the case of service development or in the more complex case of system development, which takes into account simultaneously different products and services along with the stakeholder configuration – their interactions really.

As the following chapters will show us, these methods and tools have been developed for solving three specific objectives and support the designer:

- In assessing the existing system and identifying the priorities for design intervention
- In orientating the design decisions towards greater sustainability
- In estimating the possible improvements, in sustainability terms, of ongoing development

These methods and tools have been developed (or are being developed) with the objective of being integrated in the design process, to be adapted to (or adapt) the different stages of the development process. In these chapters we will see what has been done, what are the limitations and what still has to be done (to become available for the designers). But first, a couple of observations.

It is probably obvious, but necessary to know, that to integrate the sustainability requirements – the appropriate methods and tools – during the primary development stages becomes more effective (Fig. 11.1). It is exactly during these stages that some more influential decisions are determined. The obtainable degree of innovation would be greater and so would be the potential for reducing environmental impact.

Concerning the demand for developing such instruments that are case-specific (according to the producer or product), it could be said that the methods, tools and general information on the design for environmental sustainability have their im-

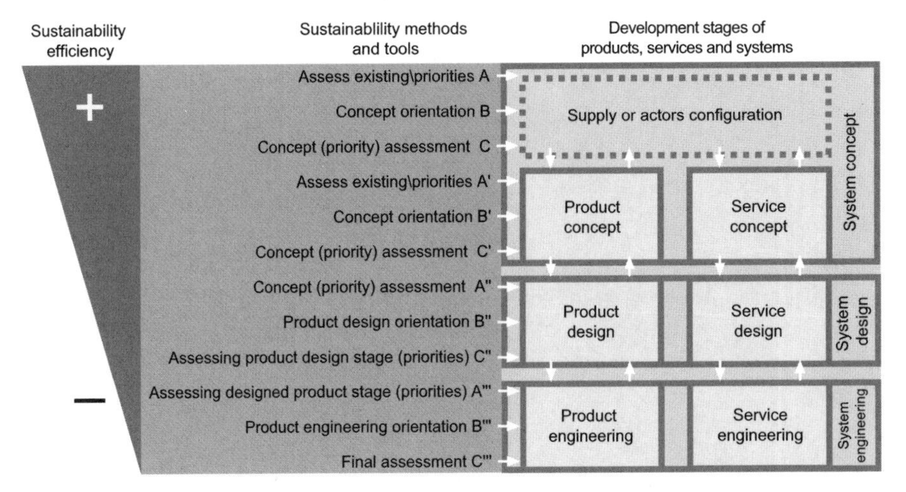

Fig. 11.1 Development stages and related methods and tools for environmental sustainability

portance on a theoretical level, but might become more efficient if they are transformed according to specific design *contexts*. Here, *contexts* refer to both commodity sectors and socio-economic/environmental characteristics of a certain space/locality.

In fact, we can observe that crosswise to the dimensions we have discussed, the discipline is specifying down to different product sectors, with the results of experience and tools in the fashion, furniture or packaging sector.

Some studies[1] have pointed out that in order to define the design priorities the specificity of certain products has to be seen through the requirements, functionality and characteristics, and on the other hand, through the profile of its environmental impact.

[1] *Cf.* Dahlström (1999)

Chapter 12
Estimating the Environmental Impact of Products: Life Cycle Assessment

12.1 The Environmental Impact of Our Production–Consumption System

Every human activity determines an uptake and acquisition of natural resources as well as release of different emissions, that is, chemical or physical agents such as various substances, noise *etc.*

Both acquisitions and emissions are forms of environmental impact. Emissions entail the release *of substances into nature*, while consumption of primary resources determines extraction *of substances from nature.* Thus, every form of impact is based on an *exchange of substances* between the environment and the production–consumption system.

This system, involving the entire complex of human activities, has and will determine an unsustainable rate of loading from and unloading into the environment, meaning that the amount of impact determined by emissions and the extraction of resources is greater than the environment can absorb without losing the overall equilibrium, also jeopardising the survival of both flora and fauna as well of humanity itself.

The impact can have different geographical amplitudes, and the problems can be listed thus:

- *On a local level*, when the effects occur in the immediate vicinity of the production site, route or landfill site
- *On a regional level*, when effects permeate across a certain geographical area, *e. g.* smog effused from industrial zones
- *On a global level*, for example, climate change

Here, we also list the most relevant environmental effects that are determined by the impacts of extraction or emissions (*cf.* compendious tables in Appendix 2):

- Exhaustion of natural resources
- Global warming
- Ozone layer depletion

- Smog
- Acidification
- Eutrophication
- Toxic air, water and ground pollution
- Landfill sites

12.1.1 Exhaustion of Natural Resources

The remaining world reserves of fossil fuels, uranium and some other resources[1] give rise to serious concerns. In fact, the exhaustion of natural resources (the *input* of our production–consumption system) is considered to be a severe problem with regard to maintaining our production–consumption system.

Using possibly depletable resources is also a crucial issue in sustainability discourse, that is, in the working hypothesis of a socio-productive model, that we should not compromise the survival and well-being of future generations. From this perspective the usage of renewable resources (materials and energy) is of great importance.

Actually, it would be more correct to speak about the renewability degree and associate it with the rate of human consumption. In fact, given a couple of millennia, all resources are renewable. Thus, the renewability is understood according to the consumption, replenishing and human demand rate. In the end, we can still claim that renewable resources are not depletable, but we have to calculate how to not compromise the natural mechanisms that are behind it.

Besides, the potentiality of renewable sources today is rather limited, for example, employing wind, hydroelectric or solar power still presents many economic and technological obstacles.

Finally, the equation – renewable resources are clean resources – is not always valid; proper assessment criteria also have to take into account the environmental impact of the extraction and enabling access to the resources. It is enough to think about steering the rivers and deaths caused by hydroelectric dams at various times and in various places of the world. Or furthermore, the possible pollution caused by processing in order to produce biodegradable polymers.

12.1.2 Global Warming

Global temperatures depend on the balance between absorbed solar radiation and infrared radiation released from Earth. The properties of some gases (carbon dioxide [CO_2], chlorofluorocarbons [CFC], methane [CH_4], nitrogen oxides [NOx {N_2O}],

---

[1] Theoretically, we can not run out of metals as they are always recyclable. Although actual exhaustion of mineral resources would have an heavy economic impact.

ozone [O_3] *etc.*) have the capacities to entrap infra-red radiation coming from the Sun to Earth. The so-called *greenhouse effect* process keeps the global temperature at the level necessary for survival of life on Earth.

Human activities have raised the appearance of such gases to disturbing levels and nowadays the *greenhouse effect* affects the thermal balance of our planet and could determine further climate changes, particularly the increase in the global temperature of the Earth.

Though the working principles of the greenhouse effect are still being studied, the predictions so far promise immense consequences: melting of the polar ice-caps, raising water levels and flooding lower areas, desertification and migration of pathogens from tropical areas.

The contribution of CO_2 to the greenhouse effect is more than 50%, CFC gases remain at 20% bar and the rest is caused by other gases. On the other hand, while the atmospheric absorption time of CH_4 is under 20 years, then in the case of CO_2, CFCs and NO_2, it exceeds a century.

The combustion of gasoline, coal and natural gas (fossil fuels) releases all the main elements that are responsible for the greenhouse effect. Additionally, the warming is caused by the traffic, house heating and power stations (thus, also by domestic or office energy consumption).

Eighty percent of CO_2 is effectuated by different processes of energy transformation (especially in the case of gasoline or coal), about 17% is caused by industrial processes and the rest (3%) by deforestation. In the case of fire-induced deforestation the emitting of CO_2 is caused both directly – as a result of burning – and indirectly – due to a decrease in the chlorophyllic photosynthesis and consequently to smaller CO_2 consumption.

Agriculture also plays its part due to fertiliser (N_2O) consumption and the CH_4 emissions caused by cattle breeding. Thus, eating meat indirectly contributes to global warming.

In the end, using a car also increases the CO_2 in the atmosphere.

The turning point in climate politics was during COP-3 (The Third Conference of the Parties) held in 1997 in Kyoto, with the signing of the *Kyoto Protocol*. The Protocol states that by 2008–2012, the industrialised countries have to have reduced their greenhouse gas emissions by 5% below the global levels on the basis of 1990. For reaching the objectives the Protocol allows industrialised countries to offset CO_2 in forests and agricultural territories (so-called carbon sinks) and to cooperate on an international level in both the forest industry and agriculture.

12.1.3 Ozone Layer Depletion

Although ozone is a toxic substance, it has the incredibly important task (for us, living) of absorbing, high up from the surface of the earth's crust, ultraviolet radiation, which is highly hazardous for flora and fauna. For humans the main problem is the increased risk of skin tumours. Current estimations show that the ozone layer is about 5–10 times thinner compared with the first estimations of the ozone layer. The main agent behind these damages is CFC, but also HCFC and tetrachloromethane. These substances, when they reach the stratosphere, provoke the transformation of molecular ozone and cause the depletion of the ozone layer. It should be added that it takes approximately 20 years for these substances to reach the stratosphere, which means that even if we stop all CFC emissions now, the damages will still occur for many years.

Direct emissions of these gases are caused by the use of the following products: sprays that contain CFC (phased out in many countries nowadays), chlorinated detergents for dry cleaning and solvent-based varnishes. Refrigerating systems (and thus indirectly also buying imported frozen food) and air conditioning also play their part. Not forgetting the polymer foaming processes that include CFC (and indirectly use of the foamed products).

The latest estimations show that supersonic planes and an increase in traffic with a tropospheric flight altitude contribute between 5 and 12% of ozone layer diminution (NOx emissions).

12.1.4 Smog

Two types of smog are distinguished: photochemical, the so-called summer smog, and winter smog.

12.1.4.1 Summer Smog

Hydrocarbons react with N_2O photolysis (sunlight) cycle and provoke photochemical smog, consisting of a high concentration of ozone, CO, PAN, and other Volatile Organic Compounds (VOC: aldehydes, ketones, hydrocarbons *etc.*) in the atmosphere. This smog is hazardous for humans, flora and fauna alike, and causes severe damage to crops, which has a deleterious economic effect. Some organic compounds (*e. g.* aldehydes) provoke lacrimation and irritate respiration, others (*e. g.* PAN) can have toxic effects on plants. Among the roots of this problem can be listed exhaust gases emitted by cars (NOx, CxHy); domestic heating systems (NOx); fertiliser consumption in agriculture (N_2O); and the activities of industries, refineries, power stations (NOx), and thus indirectly domestic and office consumption of gas, electricity and fossil fuels.

12.1.4.2 Winter Smog

During winter the SPM (Suspended Particulate Matter) and sulphur dioxide (SO_2) concentrations provoke potentially fatal respiration problems[2]. SO_2 emissions are

[2] Like the winter smog that caused 4,000 deaths in London during the winter of 1952.

mainly caused by car exhaust gases, industrial activities, refineries, power stations (domestic and office consumption of gas, electricity and fossil fuels), and toxic smoke emitted by unfiltered incineration.

12.1.5 Acidification

Nitric oxides (NO_2, NOx) transform in the atmosphere into nitric acid (HNO_3), and sulphur oxides (mainly SO_2, but generally SOx) transforms into sulphuric acid (H_2SO_4); in contact with rain water they turn into acid and determine cumulative acidity in the soil, water and on the roofs of buildings. Other substances that cause acidification are ammoniac (NH_3) and Volatile Organic Compounds (VOC).

Acidification curbs the re-growth of trees in urban zones, but also in forests, it determines a corrosion of monuments and buildings, contaminates ground water (loss of aquatic flora) and can cause sanitary risks (respiratory problems).

Plants, especially trees, that grow in sandy areas are affected by this phenomenon even more, because some toxic substances will enter into the solution, when a certain acidic level is reached.

Acidification causes are traceable back to some agricultural activities (ammoniac emitted by livestock manure), exhaust gases emitted by cars (SO_2, NOx, VOC), to activities of industries, refineries, power stations (SO_2, NOx, VOC), and indirectly to domestic and office consumption of gas, electricity and fossil fuels, to domestic heating systems (NOx and VOC), and to using cleaning products that contain ammonia (NH_3) and solvent-based paint (VOC).

The Helsinki Protocol sanctioned a reduction of SOx emissions to 30% compared with 1980. So far the results have been good, meaning that almost all member countries have kept to their restraint objectives, mainly thanks to the gradual lowering of sulphur content in fuel and substitution of several oil derivates.

Greater problems occur with the NOx emissions. The objectives of the Sofia protocol – stabilising emission levels at the rate of 1987 – have not been reached. Here, needed more incisive interventions are needed in power industry processes and structures, and in different ways of transportation. Street traffic alone is responsible for 46.7% of total NOx emissions.

12.1.6 Eutrophication

Phosphates (PO_4, salts of phosphoric acid), nitrates (NO_3, salts of nitric acid), nitric oxides (NOx), ammoniac (NH_3), nitrogen oxide (N_2O), and gaseous nitrogen (N_2) can cause a *hyper-fertilisation*, which is an accumulation of nutrients in soil and water. More susceptible for these processes are lakes and artificial water basins, where the relative speed of water exchange is considerably low and facilitates the accumulation of eutrophicating substances.

All this favours monocultures and determines a loss of the plants that usually grow in poorer soil. In water it appears as excessive growth of algae and loss of aquatic fauna together with the contamination of water bodies that cannot be used any longer as a water supply (lakes) or for swimming (lakes and ocean).

Eutrophication is caused by agricultural processes that involve phosphate and nitrate fertilisers. Thus, the consumption of intensively cultivated food products, as well as use of gardening fertilisers (phosphates and nitrates) also plays an indirect role.

Also, drainage and sewage water and industrial waste (nitrates and phosphates) are important carriers of eutrophicating agents. This why it is hard to find on the market detergents that contain phosphates; in fact, this is one of the few cases, where the environmentalist lobby has managed to break through on an industrial level.

Finally, the part street traffic plays in causing eutrophication (exhaust gases emitted by cars SO_2, NOx).

12.1.7 Toxic Air, Soil and Water Pollution

Many substances are hazardous to humans and ecosystems alike. The effects can be straight-forwardly lethal or appear after long years of exposure. There are, in fact, persistent toxic substances (non-degradable) that will create an effect after long-term accumulation. Sometimes, the carcinogenic or mutagenic effects are also hereditary.

The food chain can cause the accumulation of toxic agents in tissues and thus return to humans the substances they discarded into nature. Persistent toxic agents can first accumulate in water[3] or soil. Among those we can list heavy metals (mercury, lead, arsenic, cadmium, hexavalent chromium, nickel, selenium, zinc), chlorinated pesticides (DDT), chemical substances as polychlorobiphenyl (PCB) and polychlorotriphenyl (PCT), and of course oil and exhaust gases.

The phenomenon of the percolation of toxic substances that are not properly impermeable in landfill sites, discharging into the water bodies industrial and urban waste water containing toxic metals (mercury, lead, arsenic, cadmium, hexavalent chromium, nickel, selenium, zinc), oil and its derivatives, exhausted oils, and radioactive and chemical waste.

Via the so-called food chain all the toxic substances discarded into the environment can return and accumulate in the tissues. The lead for example can be acquired via polluted food (or directly via inhalation) and would cause poisoning (saturnism: irreversible neurological damage); similarly, the derivatives of mer-

[3] Typical indexes of pollution degree are the *Biochemical Oxygen Demand (BOD)* and *Chemical Oxygen Demand* (COD). BOD is a parameter for estimating the biodegradable organic pollution of lakes and water basins; it is equal to the amount of oxygen necessary for decomposing biodegradable pollutants, usually organic substances coming from waste. COD measures the pollutant load in a flow that might extract oxygen from the water.

cury (alkylmercury) can be found in fish products or in animals that have been fed with contaminated food. Great problems also arise from agriculture, where sometimes mercury (Hg)-containing fungicides are used.

Worrying effects are caused by the pollution of ground water levels, as this makes the water non-drinkable and non-usable for irrigation.

Buying products equipped with batteries (toys, equipment, domestic appliances), or with insulation, transformers and condensers (containing PCB); thermometers and manometers (Hg), that probably will end up in the landfill site, are the indirect causes of soil pollution.

Among toxic substances spread by air we should list the fumigant insecticides (pesticides), carcinogenic aromatic hydrocarbons (pyrene, benzopyrene, benzene), particles of asbestos, beryllium, lead, mercury, chromium, vinyl chloride and dioxane.

Incineration without proper filtering systems can emit toxic gases and smoke (SO_2 and dioxane); dioxane (TCDD) provokes chloracne and soft tissue cancer.

The combustion of biofuel (which contains benzene) without catalytic mufflers, but also burning tobacco, that is, the smoke from a cigarette, produces pyrene and benzopyrene. When inhaled, these substances prove to be highly carcinogenic. Among other reasons, using a car will become truly hazardous when it consumes lead-containing petrol (Pb emissions).

Polychlorobiphenyl (PCB)-containing pesticides are especially dangerous, because PCB has toxic effects on brain and liver tissue.

Volatile Organic Compounds (VOC) containing solvents can cause emissions and become pollution agents throughout their life-cycle.

Finally, we should not neglect the damage caused by radioactivity, either due to nuclear explosions or to incidents in nuclear power stations: immediate bodily damage (nausea, blood modifications, infections, death), delayed bodily damage (leukaemia, tumours), genetic damage (mutations in progeny).

12.1.8 Waste

Many countries have to face the problems of a lack of appropriate places for landfill sites, soil and ground water pollution, olfactory pollution and explosion dangers in landfills, as well as problems with waste transportation (fuel consumption, noise, air pollution).

In landfills that are not properly controlled, as we mentioned before, due to humidity the heavy metals can leak from batteries into nature and permanently pollute ground water levels. Batteries and accumulators are greatly responsible for the presence of mercury, cadmium, zinc and nickel in landfills.

All this is strictly correlated with producer strategies as well as with the behaviour of consumers. In fact, a consumer can also make choices according to packaging use (single or multiple) or according to the characteristics of product engagement (throw-away or durable goods). Consumer behaviour can also determine premature disposal, for example in the case of cultural obsolescence.

In the end, we can be (culpably) careless in searching for possibilities of secondary use, re-selling or donating goods, or just because we try to avoid the small effort of waste selection.

12.1.9 Other Effects

There are other connected phenomena like olfactory pollution by the production–consumption system, acoustic pollution and degradation of the landscape. In the case of the latter phenomenon, more than the former, it is remarkably hard to calculate the exact value of such processes.

12.2 Quantitative Methods for Estimating and Analysing Product Environmental Impact

The quantitative methodologies for analysing: and estimating the environmental impact are seeking to analyse, estimate and interpret the relations between the product and the environment.

For three reasons these calculations prove to be rather complex.

First, it is not about the product and even less about the materials, that determine the environmental impact, but the sum of processes that occur during the life cycle. This is why it is necessary to prepare a life cycle model, from the extraction of raw materials to the disposal processes. This modelling can raise many uncertainties[4].

Second, even when the entire life cycle is profiled, many actual impacts of processes remain doubtful. Despite great progress otherwise, there is still a considerable lack of reliable data.

Finally, understanding of the environment that surrounds us is still limited. Nature is extremely complex and intricate to model. Also, cause–effect relationships are hard to isolate.

Summarily, the methodologies of life cycle analysis raise two kinds of criticisms:

- Models are too complex (expensive) and not usable in industry
- Methods simplify the actual situation too much and become unreliable from a scientific perspective

In fact, the analysis of life cycles is always more or less about compromise between practicality and complexity. The depth of complexity is usually chosen according to the actual goal of the project.

[4] We know little, for example, about user behaviours, and similarly hard to calculate are the end-of-life treatments.

The *Life Cycle Assessment (LCA)* is definitely the best-established[5] methodology and better than others in the above-mentioned problems.

12.2.1 Life Cycle Assessment

It has been said that *the life cycle* concept refers to the sum of interactions that a product has with the environment, consisting of the pre-production, production, distribution, use, re-use, maintenance, recycling and end-of-life treatment processes.

First, an institution of this kind in history that dealt with various experience gained internationally in the field of development objectives and LCA terminology was SETAC[6]. Today, LCA has found official recognition in international standards being introduced to ISO (International Standardisation Organisation) normatives.

According to the ISO 14040 definition, LCA is a technique estimating the environmental aspects and potential long-term impacts of the whole life-cycle of products or services, namely:

- Compiling and inventorying the implications of the system's inputs and outputs
- Evaluation of the potential impacts regarding these inputs and outputs
- Interpreting the results of inventory and evaluation phases according to given scope and objectives

Life Cycle Assessment considers the ecological impacts of the systems examined, particularly in the light of environmental and human health and depletion of natural resources, but does not observe its economic and social character.

One has to remember that this methodology works with models; therefore, it is a simplification of the real world and does not pretend to handle environmental interactions in an absolute and unerring manner.

General objectives elaborating LCA are:

- Defining the framework of interactions between a given activity and the environment as integrally as possible.
- Contributing towards further understanding of the complexities occurring with the environmental impacts of such activities.
- Provide all interested sides who have any power on further decisions[7] with information about the impacts of such activities on the environment and about opportunities to improve environmental conditions.

---

[5] Another sophisticated methodology is MIPS (Material Input Per Service), which was developed by the Wuppertal Institute, Germany, as an efficient evaluation and a round-up comparison tool for measuring the intensity of environmental impact on different products and services.

[6] SETAC is the acronym for the Society for Environmental Technology and Chemistry.

[7] Referring to either authorities who define the normatives or, for example, those who attribute the eco-label, or those who during different stages of product development are in a position to make decisions.

12.2.2 Stages of LCA

The elaboration process of LCA is divided into four[8] respective phases (Fig. 12.1):

- Goal and scope definition
- Life Cycle Inventory
- Life Cycle Impact Assessment
- Interpretation of the results

12.2.2.1 The Goal and Scope Definition

This stage of LCA has been well defined on a theoretical level and is consequently divided into four steps.

Definition of research goal – identifying the reasons behind LCA development and the engagement of the results, *e. g.* inside or outside the company.

Definition of the scope – identifying the production system, its reach and limits, during later stages, it is opportune to return to the definition of the scope.

Definition of the functional unit – one of the crucial steps in LCA, where the dimensions and assessments the system studied has to undergo are defined. In other words, it is not the product to be assessed, but all processes associated with the fulfilment of a given function. LCA can be applied to both physical and immaterial products, *e. g.* services.

Comparing possible alternatives, for example, the product before and after re-design, has to match functionally similar products, services and processes. The *functional unit* is defined in order to become the basis for such comparisons[9].

Definition of data quality – minimum criteria for using available information[10].

12.2.2.2 Life Cycle Inventory

Life Cycle Inventory has been also well defined, but requires some clarification.

During the inventory, the output and input of the system studied is analysed against the background of the functional unit defined within the goal and scope definition. After the *definition of system boundaries* and the composition of a *process flow-chart,* the next step is data processing. This phase can be divided into:

---

[8] ISO 14000.

[9] *E.g.* transportation per persons: the functional unit is *passenger per kilometre*. If all transportation systems are being examined, then the comparisons, for example, are made between a bicycle, motorbike, car, public transport, train and plane; eventually several restrictions can be applied, as in the case of private transportation, trains and planes can be ignored, and reducing street traffic does not have to take into account trains, planes and bicycles. In all cases the life-span of products/services and the re-use and recycling possibilities have to be compared.

[10] Here, it is decided, for example, if the data has to be surveyed or whether it is possible to use the information *as it is*.

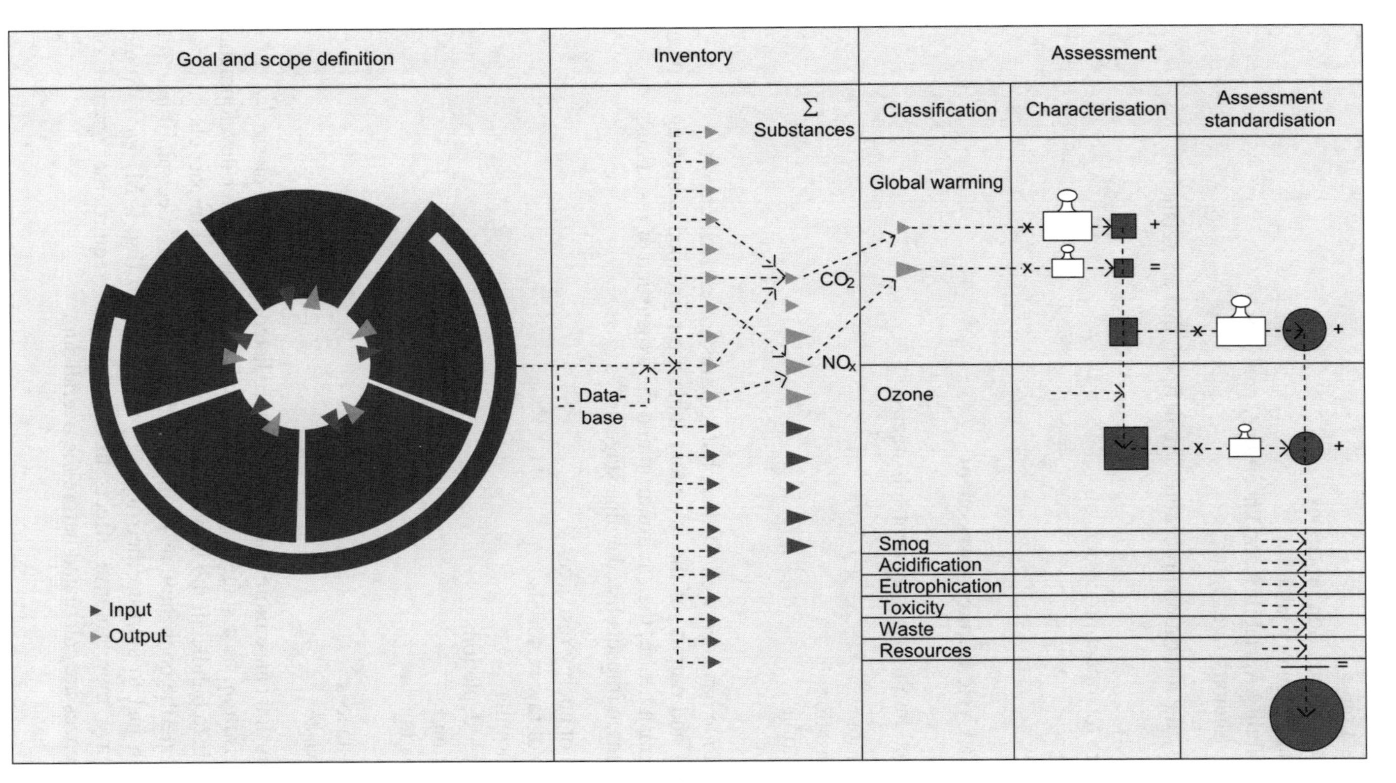

Fig. 12.1 Life Cycle Assessment stages

- Data collection
- Defining the calculation processes
- Creating an inventory table
- Data sensibility, variability and uncertainty analysis
- Defining voluntary oversights

And finally the allocation procedures are defined, that are related to:

- Co-products
- End-of-life treatment processes
- Recycling

12.2.2.3 Life Cycle Impact Assessment

This is divided into four subsequent sub-stages:

- Classification
- Characterisation
- Normalisation
- Evaluation

Classification has been well defined, but requires some clarification.

All inputs and outputs of the inventory table are regrouped according to their impact on human health, the environment and the exhaustion of resources.

Most common impacts listed at this stage are[11]:

- Depletion of energy resources
- Depletion of raw materials
- Global warming
- Ozone layer depletion
- Acidification
- Eutrophication
- Smog
- Toxic substances
- Polluted waste

Substances that cause more than one effect are listed in different classes[12].

Characterisation, as a stage, has been defined, but is still under development. Its objective is to bundle all the impacts into one environmental impact class.

To aggregate the contributions of input and output towards a certain impact, it is not enough just to sum up their units of measurement (kg, l, MJ *etc.*). Some substances have more intense effects than others and require that greater or smaller influences are calculated before the overall sum.

---

[11] *Cf.* Chap. 11 and Appendix 2.

[12] *E.g.* NOx are, for example, toxic, acidic and cause eutrophication.

Practically, the contribution of all extractions and emissions towards a certain environmental effect is calculated, multiplying one by one with a certain equivalence factor[13] that shows their relative contribution:

characterised effect value (problem i) = $\sum$ (input/output) equivalence factor (problem i/output i/input k) × quantity (output i/input k)

Normalisation – all environmental *characterised effect values* are adjusted on a determined *normal* scale[14]. Practically, the value effects are divided by the normal factor:

normalised effect value (problem i) = characterised effect value (problem i) / effect factor "normal" (space, time, problem i)

Life Cycle Impact Assessment – here, different contributions towards effect are calculated in order to sum their total amount[15]:

effect value (problem i) = weighting factor (problem i) × normalised effect value (problem i)

During this phase the impact information is aggregated even further; the final objective is to have a coherent result that would define the environmental impact (damage) of all impacts:

total impact = $\sum$ (problem) effect value (problem i)

This LCA stage has been conceptually defined, but currently different methods and approaches[16] are employed. These assessments can also reflect social values and decisions tend to be political rather than scientific. From the other side is hard to imagine that in the near future a strictly scientific theory would occur[17].

12.2.2.4 Interpretation

During this stage the results of the inventory and assessment phase are interpreted according to the original goals and scope of the research and can consequently bring forward conclusions and advice for whoever is in a position to decide.

This stage can also lead to revising the original goals and scope according to the character and quality of the data collected.

---

[13] *E.g.* ODP (Ozone Depletion Potential) is a parameter that measures relative potentials of different gases on ozone layer depletion.

[14] More common criteria are illustrated in Sect. 12.2.2.6.

[15] *E.g.* when comparing two systems, if one determines a smaller risk of ozone depletion and the other has fewer toxic emissions, then it is impossible to declare which one has worse environmental characteristics without calculating the relative importance of both impact categories. *Cf.* Sect. 12.2.4.

[16] Some of the examples in Sect. 12.2.2.6 illustrate the more important criteria.

[17] *Cf.* Sect. 12.2.4.

Finally, this is the first moment to integrate technological, economic, performance, cultural and social aspects with environmental issues.

12.2.2.5 Possible Implementations

Life cycle assessment can be a decision support system for a wide area of implementations. Here, various internal and external possibilities are listed.

Internal implementation; the results have not been published and are used for:

- Product, service or process design[18]
- Designing environmental development strategies
- Identifying opportunities for improving environmental performance
- Helping to make purchasing decisions
- Developing environmental auditing and minimising waste

External implementation; the results have been published. The information is no longer private and stricter standards are required for reliability and transparency. It is used for:

- Marketing
- Defining the eco-label criteria
- Public communication and education
- Decision support in the political field
- Decision support in purchasing procedures[19]

12.2.2.6 Current Methodologies Present on the Market

Here, we list some more established methodologies for the assessment stage of LCA.

The first was developed by BUWAL (Department for Environment of Switzerland) at the beginning of the 1990s (Example 12.1). Later, CML Centre at Leiden University, the Netherlands, conveyed significant studies for defining the criteria of the assessment stage. These studies were later reformulated[20] and gave birth to the Ecoindicator 95 method; the latter aggregates the environmental impact into a single total value and is one of the most popular tools in Europe.

Another important methodology, the EPS (Environmental Priority System; Example 12.3) has been developed by the Swedish Environmental Research Institute and is used mainly in Scandinavian countries.

---

[18] The possibilities and limitations of this application are defined in Chap. 13.

[19] A very well established method that still has to make up some ground is the Environmental Product Declaration (EPD).

[20] The project was supported by the Government of the Netherlands and developed by CML in cooperation with Prè Consultant, Philips, Ocè, Nedcar and Fresco.

Examples

12.1 BUWAL method

Based on the objectives of Swiss national policies, the input and output are calculated straight away, without any classification. Standardisation results from comparing the input–output level with a maximum or purpose level. The assessments are made comparing current levels with the purpose level.

In the end, different values are summed up in one result. The main obstacles are based on the fact that not many substances are measured for standardisation and assessment.

The final ecofactor is calculated according to the following input–output equation:

F = 1/f(k) × f/f(k) where f = current level f(k) sustainable level

12.2 CML and Ecoindicator method

The analysis and assessment criteria are described in the following (Prè Consultant, 1997; Guinée *et al.*, 1993a, 1993b).

The analysis and assessment pass the following stages:

Classification

Taking into consideration the following environmental problems:

- Depletion of energy resources
- Depletion of raw materials
- Global warming
- Ozone layer depletion
- Acidification
- Eutrophication
- Summer smog
- Winter smog
- Toxic substances
- Solid waste
- Heavy metals in air and water
- Carcinogenic substances
- Pesticides

Characterisation

Equivalence factors that indicate an input or output contribution, and are relevant to the specified problem are:

- Depletion of energy resources [MJ] = Σ (energy) input of energy (energy i) [MJ]
- Depletion of raw materials = Σ material consumption (input i)[kg]/reserves (input i) [kg]

- Global warming [kg] = Σ (output) GWP (output i) × output (output i) in air [kg], where GWP = Global Warming Potential
- Ozone layer depletion [kg] = Σ (output) ODP (output i) × output (output i) [kg], where ODP = Ozone Depletion Potential
- Acidification [kg] = Σ (output) AP (output i) × output (output i) in air [kg], where AP = Acidification Potential
- Eutrophication [kg] = Σ (output) NP (output i) × output (output i) [kg], where NP = Nutrification Potential
- Summer smog [kg] = Σ (output) POCP (output i) × output (output i) in air [kg] where POCP = Photochemical Ozone Creation Potential
- Winter smog [kg SO2 or SPM] = output of SO_2 in air [kg] + output of SPM in air [kg]
- Solid waste Σ (output) solid waste (output i) [kg]
- Heavy metals in air and water [kg Pb] = Σ (output) AQG × output (output i) in air [kg]/AQG(output i) + GDWQ (Pb) × output (output j) in water [kg]/GDWQ (output j), where AQG = Air Quality Guidelines, GDWQ = Guidelines for Ground Water Quality
- Carcinogenic substances [kg Pb] = Σ (output) AQG × output (output i) in air [kg]/AQG(output i) where AQG = Air Quality Guidelines
- Pesticides [kg] = Σ (output) active indicators (output i)

Normalisation

The environmental impact created by an average European in a certain time span is taken as a measure of a normal profile. Practically, the score of any effect can be divided with a relatively normal effect:

normalised effect value (problem i) = effect value character (problem i) / normalised effect factor (space, time, problem i)

Life Cycle Impact Assessment

Normalised effect values are judged according to the distance from purpose; basically, the severity of a problem is evaluated on the basis of the difference between current and purpose levels. Results are added up by environmental effects and can be presented as just one aggregated value.

The purpose level is defined as follows:

- At the purpose level the effect will cause no more than one death per million inhabitants a year.
- At the purpose level the effect will damage or jeopardise less than 5% of ecosystems in Europe.
- At the purpose level the smog periods are a lot less probable.

This means that the death per million inhabitants is taken as being severe, damaging or as jeopardising 5% of European ecosystems.

Judgement factors:

Greenhouse effect – some studies show that the greenhouse effect would damage or jeopardise less than 5% of ecosystems, if reduced by a factor of 2.5.

Ozone layer depletion – the Montreal Protocol has agreed to eliminate all CFC emissions. The HCFC emissions are reduced until complete phasing-out is achieved in 2015. It has been calculated that in this case the effect will not exceed the 1 death per million inhabitants a year bar. It has also been calculated that such a reduction would diminish the ozone layer depletion to 1% of its current level. The reduction factor is considered to be 100.

Acidification – some studies show that the acidification will damage or jeopardise less than 5% of ecosystems, if reduced to 5–10% of its current level. The reduction factor is considered to be 10.

Eutrophication – in some rivers and lakes the purpose level is exceeded 5 times. Emissions have to be reduced by a factor of 5.

Summer smog – some studies show that in order to prevent smog periods and health complaints, particularly among the elderly and people who suffer from asthma, the effect has to be reduced by a factor of 2.5.

Winter smog – some studies show that in order to prevent smog periods and health complaints, particularly among the elderly and people who suffer from asthma, the effect has to be reduced by a factor of 5.

Heavy metals in water and air – a reduction by a factor of 5 has been agreed.

Carcinogenic substances – it has been calculated that the effect will not surpass the 1 death per million inhabitants a year, if reduced by a factor of 5.

Pesticides – a European standard has established a contamination limit; this has been exceeded in 25% of cases. A reduction factor of 25.

Solid waste – there has been no agreement on reduction as the effect is not lethal and only a small proportion of ecosystems are damaged by landfill sites. Greater problems are connected to the emissions by incineration, the composting of waste and heavy metal contaminants. This waste is considered and evaluated according to its own LCA.

Depletion of energy resources – no reduction factor has been agreed on for the depletion of energy resources. Greater problems are connected with the consumption of fossil fuels; this has been calculated as part of other environmental issues. Alternatives for substituting fossil fuels in energy transformation exist, provided that the market is ready to pay higher prices.

Depletion of raw materials – no reduction factor has been agreed on, because no deaths or ecosystem damages can be attributed to the depletion of raw materials. The problems here are first and foremost social and economic.

Ecoindicator

The Ecoindicator gives a single value for a environmental impact, adding up the assessment results:

Ecoindicator = $\sum$ (problem i) judgement factor (problem i) $\times$ standardised effect value (problem i)

12.3 Environmental Priority System method

The Environmental Priority System (EPS) method translates the environmental impacts into a sort of social expenditure.

In order to calculate the social costs, the damage inflicted on subjects with community value is identified. Then, how much the society is ready to pay for such subjects is identified. The resulting amount (ELU) are added to give a single value. Many assessment principles are applied for these calculations.

12.2.3 LCA and Design: Importance and Limitations

12.2.3.1 Importance

As has been said, LCA is the most reliable method for calculating environmental impact to date.

Interests in this method are growing, but several nodal points have to be resolved first, especially when it is considered as an auxiliary tool for design. In fact, an analysis and assessment methodology, like LCA, is not literally a support tool for product development. Besides, such a tool should not only indicate when something is better than others, but also direct the user towards possible solutions and strategies.

Here, we are going to see what the limitations are, in general and in detail, of this tool with regard to design activities.

12.2.3.2 General Limitations of LCA

Due to the complexity of the relationships that LCA analyses, several problems arise:

- LCA assumptions and choices might be subjective.
- Models used for inventory analysis and impact assessment are not sufficient to describe the entire spectrum of environmental impacts, nor can they be adapted to all applications.
- The results and criteria of global LCA might not be adaptable for local applications.
- The reliability of LCA results might be limited due to missing, unavailable or low quality data.
- Problems with implementations often arise due to a lack of information regarding certain processes relating to specific products. Public or low cost databases do not include data on all processes, especially if they happen to be less common.

12.2.4 Power to Choose: Discriminant Power Versus Scientific Reliability

A very important moment for understanding the impact of a certain product is the assessment of an inventory table, as it gives information that can be analysed and utilised for decision-making (Fig. 12.2).

For whoever is in charge, the optimal format would be a single assessment (absolute aggregation), based on all possible effects and proceeding from valuations as scientific as possible.

In reality, however, the reliability of results is in inverse proportion to their aggregation, or discriminant power. In the course of an LCA it often happens that:

- Increasing the number of different impact categories will decrease the capacity of defining credible aggregation methods.
- Increasing the aggregation depth will decrease the scientific quality of the assessment and increase the relative importance of political or social judgements.
- Increasing the aggregation depth will increase the discretionary capacity to choose alternatives with smaller impact.

This does not mean we should avoid making decisions, just because they have not been scientifically ascertained; in the end, we are also political beings, not only scientists. In sustainable development, the risks are minimised, but some decisions still have to be made.

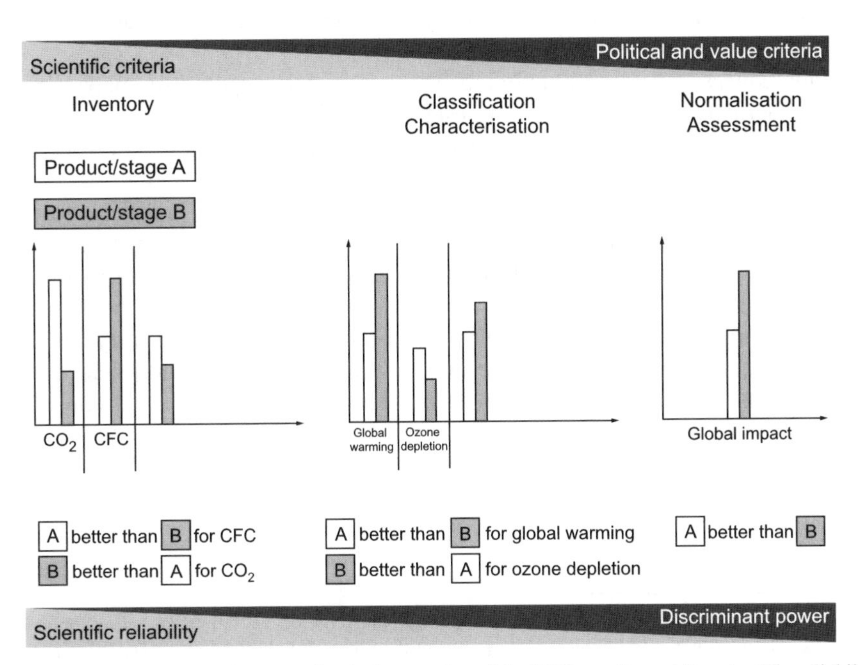

Fig. 12.2 Correlation between discriminant value of the LCA results and its scientific reliability

12.2.5 Incisive Decisions: First Stages of Development Versus LCA Applicability

Improving the product impact is more probable during the initial stages of development, when the innovation has greater magnitude.

Unfortunately, in the course of product development, the more everything becomes defined, the greater the possible success of an LCA development. This is because the LCA requires greater quantities of information, which are impossible to have during the initial stages (brief/product strategy and concept design).

More evident reasons that prevent the application of LCA during the initial stages are:

- The design process goes through a great amount of ideas and alternative choices; many are discarded, assessments have to be made fast and often there is no time for full LCA on every subject.
- A product concept is not a product; it is hard to proceed with an LCA that requires precise, quantitative information.
- The indications during these stages are derived mainly from evaluation rather than from an inventory table; evaluation at the more critical stage.

During the initial stages of design, LCA is usually used only for analysing existing products and services. Theoretically, we can guess which stages and processes will have a greater impact and formulate the environmental priorities of the product. Proper identification of environmental priorities is crucial for guiding design efforts and eventually establishing the selection criteria for alternative solutions[21].

During *concept design,* the environmental assessments require fast analysis of possible alternatives for development design (what-if analysis), in order to compare different solutions. Complete LCA might prove unadaptable or too taxing to be carried out.

During *product design*, the environmental assessments require preliminary estimates from fast and simplified applications of LCA for identification of critical impact factors. Even when carried out, they have to be rechecked and verified later.

Finally, during the *engineering*, the environmental assessments can carry out, if necessary data are available, full-scale LCA on the technological and production options. Also, final verifications of objective (reduction of environmental impact) fulfilment can be undertaken.

[21] *E.g.* LCA on a washing machine would assume that the stage of utilisation has a greater impact; thus, the priority would be to consider reducing the consumption during use (energy, water and detergents) and design strategies should follow this lead and try to solve the problems.

12.2.6 Developing LCA

It sounds banal, but without any data on input and output that can be analysed, there will be no assessment of environmental impact. But the data, as we saw, are far too often too generalised or decontextualised.

For this reason, many latest studies and projects, public and private, have focused on collecting reliable and specific data.

We can say that in the near future there will be information available about many materials (Example 12.4), products and/or commodity sectors, as well as about different geographical locations (Example 12.5), product, use and disposal.

Besides the data collection and specification for LCA, new methods and tools for faster or more simplified assessments (simplified LCA) are also being developed or are already on the market (Example 12.6).

These methodologies have a smaller learning curve and cost less than regular LCA; thus, they are more applicable for the initial stages of development.

Here, the problems lie in the validity and transparency of both used data and results. As these tools are more simple, they can also be used by people who are not experts of environmental analysis (that could be a bonus), but for the very same reason they can lead to erroneous assessments.

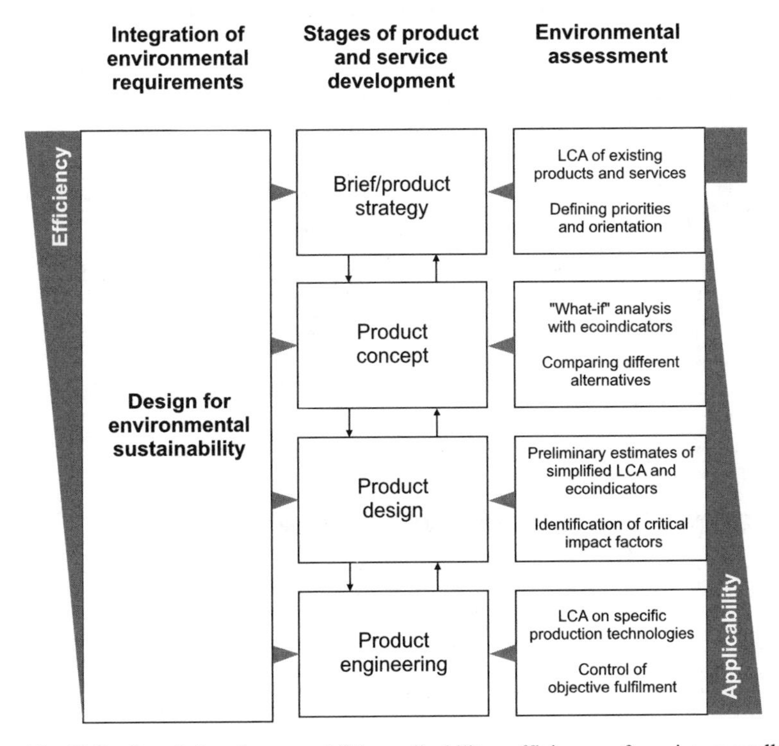

Fig. 12.3 Correlation between LCA applicability, efficiency of environmentally conscious design and stages of product development

Finally, it should always be remembered that these methods, even with reliable data, restrict their field to analysis. That is, they do not offer any guidelines for design. While they point out where the problems are, they do not give any suggestions or solutions, which leads us to the topic of next chapter.

Examples

12.4 PWMI database of plastic materials

The Association of Plastics Manufacturers in Europe (APME) has for several years financed studies and research that have defined or will define a growing number of inventory tables for polymer production. Information is collected on a European level and presented as a valid average for the member states of the EU.

12.5 Italian Database for LCA (I.LCA)

APAT (Agenzia per la Protezione dell'Ambiente e per i Servizi Tecnici, previously ANPA) has commissioned the Politecnico di Milano University to develop a (first) Italian database for LCA. This collection includes more than 400 inventory tables of various product and service processes in Italy. The database is structured into four sectors: materials and production processes, energy, transportation and end-of-life treatments. Access and utilisation is free for all companies, research institutions, educational institutions and public administration.

12.6 Eco-it

Software developed by Prè Consultant, Netherlands, that describes summarily the product life cycle and calculates its total impact.

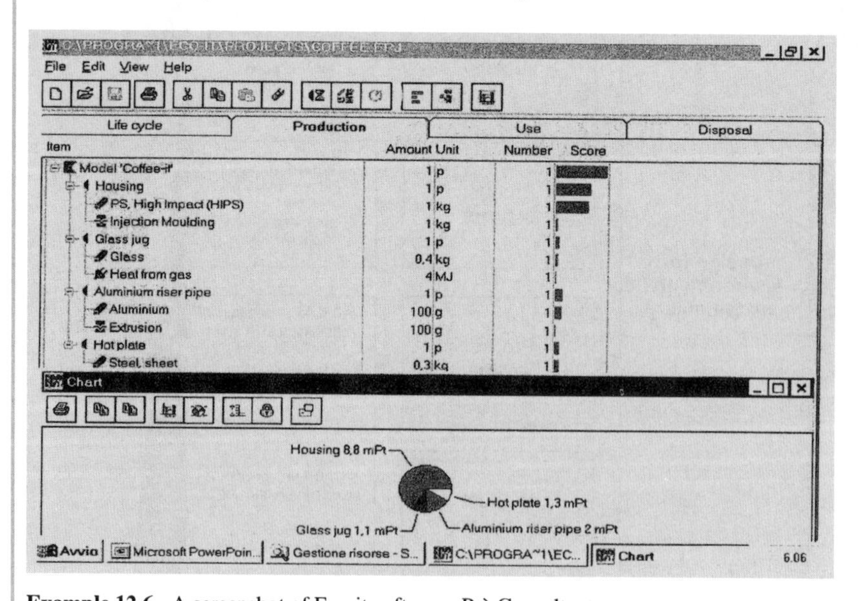

Example 12.6 A screenshot of Eco-it software, Prè Consultant

The scoring of the environmental impact is calculated for a great number of processes, proceeding from the prior information of Ecoindicator 95 (described previously). The final score is evaluated by simply multiplying by the quantity of processes that characterise the products under consideration (*e. g.* kilograms of materials required, MJ of energy consumed *etc.*).

Chapter 13
Environmentally Sustainable Design-orienting Tools

13.1 Introduction

In this chapter we speak about some tools that have been developed exactly for design process orientation towards environmentally sustainable solutions. These are divided up and described according to the different stages at which they are used;

- Tools developed for certain environmental goals
- Tools for product LCD
- Tools for design for eco-efficiency

13.2 Tools Developed for Certain Environmental Goals

Historically, the first tools were created as supplementary tools for design processes to increase certain environmental benefits. These are instruments meant for:

- Selecting low impact materials
- Minimising toxic or hazardous materials
- Designing for recycling
- Designing for disassembly
- Designing for re-manufacturing
- Different environmental standards and regulations

These tools (Examples 13.1–13.8) are usually handbooks, guidelines or support indexes either in printed or in electronic form.

Examples

13.1 Design for Plastic Recycling

Published by GE Plastic, USA, it is a manual with strategic guidelines for assisting the designer while designing for recycling.

13.2 MMU Indexes

Published by Manchester Metropolitan University, UK, these indexes are meant to keep the score on systems calculating the aspects of reducing environmental impacts on determined products.

They are easily calculable and allow *re-design* to be distinguished from the rest. The values are unified, thus allowing the final results to be shown on a standardised scale. They estimate the design improvements in recycling and disassembly. The higher the score, the better the result.

13.3 REStar

Software published by the Green Design Initiative of Carnegie Mellon University, USA, for supporting design for disassembly, recycling and repairs.

13.4 DFE (Design for Environment)

Software published by Boothroyd, Dewhurst Inc., USA, and TNO (Institute of Industrial Technology), Netherlands, for analysing disassembly of products and possible profits effectuated by recycling optimisation.

13.5 RONDA (Recycling Orientated Database Analysis)

Software published by IPA, Frankfurt, that provides information on products according to their recyclability and offers a special database for such analysis for the producers.

13.6 RECREATION (RECycling REsources And Technologies InformatiON)

Database published by IPA, Frankfurt, including all information necessary for recycling and about suppliers of recycled materials.

13.7 RECOVERY (REmanufacturing Cost Optimization Via Extended Reuse and disassemblY)

Software published by IPA, Frankfurt, which helps to find economically optimal strategies for re-manufacturing through disassembly assessment.

13.8 IDEmat

Software by TU of Delft, Netherlands, for selecting low impact materials. Contains a database on the physical, mechanical, and environmental characteristics of various materials.

Gives the possibility of searching for materials according to requirements and characteristics wanted. Environmental data conveyed by Ecoindicator and EPS.

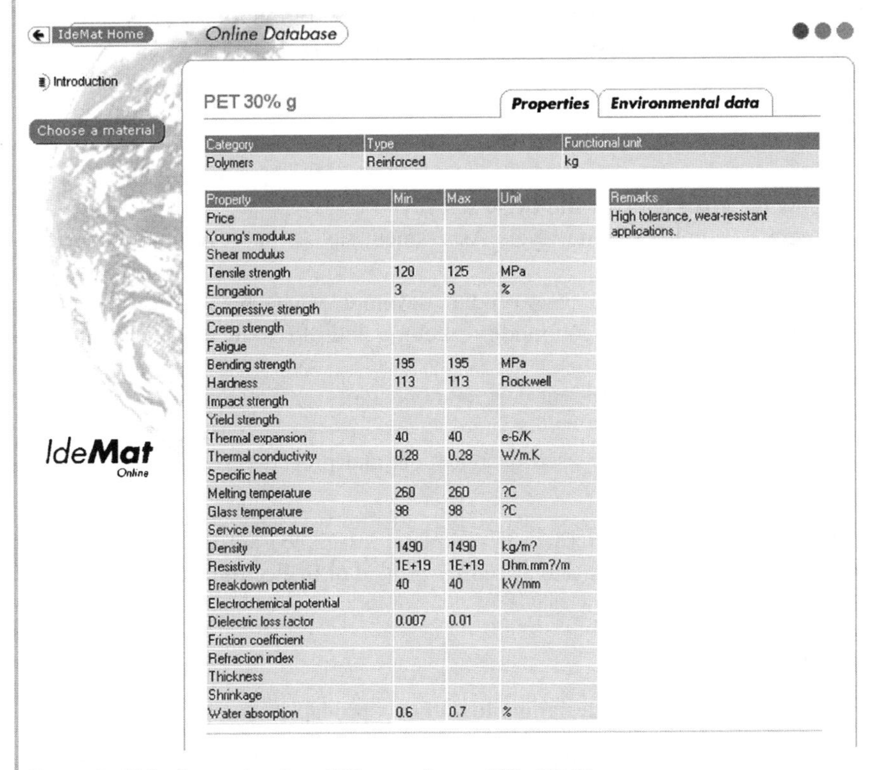

Category	Type	Functional unit
Polymers	Reinforced	kg

Property	Min	Max	Unit
Price			
Young's modulus			
Shear modulus			
Tensile strength	120	125	MPa
Elongation	3	3	%
Compressive strength			
Creep strength			
Fatigue			
Bending strength	195	195	MPa
Hardness	113	113	Rockwell
Impact strength			
Yield strength			
Thermal expansion	40	40	e-6/K
Thermal conductivity	0.28	0.28	W/m.K
Specific heat			
Melting temperature	260	260	?C
Glass temperature	98	98	?C
Service temperature			
Density	1490	1490	kg/m?
Resistivity	1E+19	1E+19	Ohm.mm?/m
Breakdown potential	40	40	kV/mm
Electrochemical potential			
Dielectric loss factor	0.007	0.01	
Friction coefficient			
Refraction index			
Thickness			
Shrinkage			
Water absorption	0.6	0.7	%

Remarks
High tolerance, wear-resistant applications.

Example 13.8 Screenshot from IDEmat software, TU of Delft

13.3 Limitations of Tools that Are Developed for Certain Environmental Goals

These tools can be useful, but they still are just supporting the solutions they were developed for. A more appropriate working hypothesis should take into account the Life Cycle Design (LCD) approach. It can happen that sometimes these tools, in order to for example minimise the impacts, actually neglect more important problems (or stages) within the same product system, or end up in difficulties when integrating the results into the process and other designing tools.

Nevertheless, they are important instruments; similarly important are the links that connect them with impact assessment systems and use LCD logic.

13.4 Tools for Product LCD

Different tools for supporting design have evolved (and are evolving), amplifying their capacities and efficiency within the framework of minimising environmental impact throughout the entire product's life cycle.

Some guidelines could be outlined here:

- Apply the goals at every life cycle stage.
- Integrate (simplified) Life Cycle Assessment (LCA) into every product development stage.
- Integrate guidelines into every product development stage.
- Specify according to commodity sectors and different product/by-product types.
- Adjust according to different development stages and involved specialists (*e. g.* new interfaces).
- Integrate with currently used design tools and procedures (such as CAD and CAM).

These tools for product LCD (Examples 13.9–13.13) are usually manuals or guidelines, support indexes, and are either printed or in electronic form.

Examples

13.9 ICS tool-kit for designing low impact products

This is a practical and simple software tool-kit configured to cooperate with the designer in the design of low environmental impact products. The tool-kit is structured with a set of idea tables containing guidelines (aiming at sustainability-focused idea generation) and a set of corresponding checklists (control questions that verify if the design has been done properly as stated in the sustainability criteria).

The criteria, the guidelines and the checklists have been developed by the Design and Innovation for Sustainability research unit of the Politecnico di Milano University, and correspond to those presented here in the Part II. This tool is freely available at www.lens.polimi.it.

13.10 Ecodesign. A Promising Approach to Sustainable Production and Consumption

Developed by TU Delft, Netherlands, by commission of UNEP, it provides a global approach to defining strategies for developing sustainable products, and a pragmatic methodology for integrating environmental requirements into normal design practices.

13.11 EDIP (Environmental Design Strategies, Environmental Specification, Environmental Design and Rules)

Developed by TU Denmark, this is a set of software for supporting design decisions during various development stages, integrated with an LCA system.

13.12 ECODesign Tool

A computer program developed by the Design for the Environment Research Group-MMU and Nortel, UK, based on expert rules for assisting design decisions during various development stages. Integrated with CAD. This system is created within a greater project of Design for Environment Decision Support (DEEDS) and was set in motion by the Engineering and Physical Sciences Research Council.

13.13 E.IM.E tool

Software developed in cooperation among Ecobilance, Association of Electronic Industries and an alliance of companies. Made for the electronics sector, this tool is efficient at many life cycle stages and for different specialists (various interfaces).

13.5 Tools for Design for Eco-efficiency

Developed methodologies and tools for design for eco-efficiency are still very limited (Example 13.14). Critical issues in such development are always connected with the difficulties of analysing complex systems. The criteria, however, should not be different from the product level, apart from placing greater emphasis on reductions in transportation and distribution, as these stages are more visible on a system level. Meanwhile, the guidelines are essentially different in the light of possible design decisions: they really should orientate towards promising interactions (configurations) between the *satisfaction system* stakeholders.

Example

13.14 Sustainability Design-Orientating (SDO) Tool-kit

The Sustainability Design-Orientating (SDO) Tool-kit was developed within the MEPSS (Methodology for Product Service System) research and was financed by the VI Framework program of the EU. The goal of this tool-kit is to orientate the design of (product service) systems towards sustainable solutions. To achieve this, it has defined the priorities for several criteria of sustainability (and provides proper checklists).

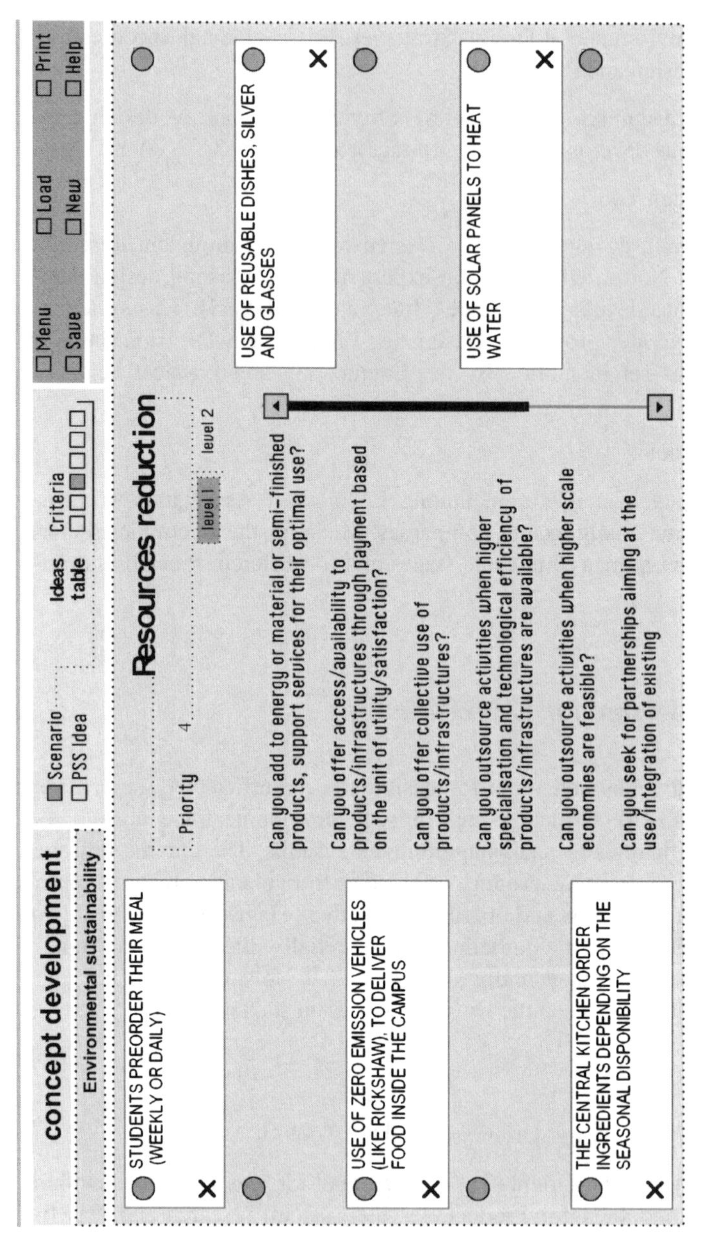

Example 13.14a Screenshot of an idea table showing the conceptual improvements

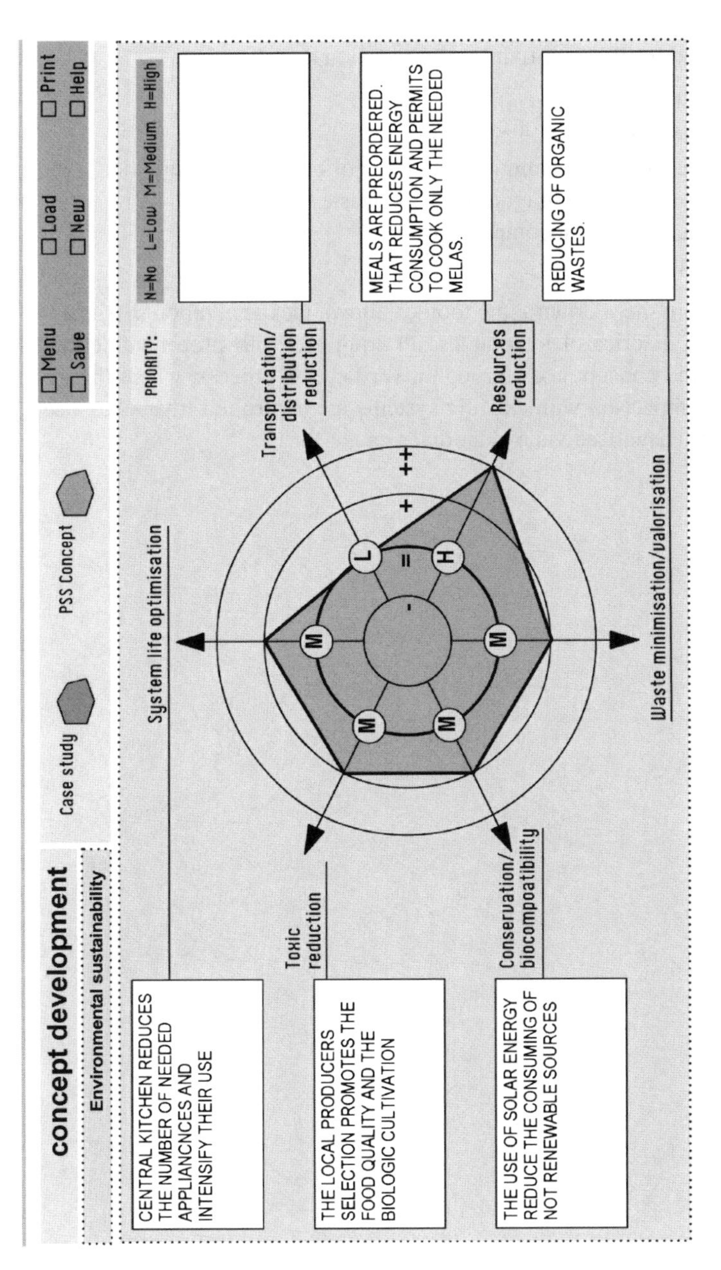

Example 13.14b Screenshot of a radar diagram showing the conceptual improvements

Criteria adapted for the environmental dimension are:

- Optimisation of system lifespan
- Reduction of transportation/ distribution
- Reduction of resource consumptionReduction of resource consumption
- Minimisation/valorisation of emissions and waste
- Improving renewability/biocompatibility
- Non-toxicity, non-hazardous

Proceeding from these criteria, the tool-kit allows idea generation sessions to be managed with a series of idea tables, all equipped with proper guidelines. Once designed, the concept goes through a verification process, where the improvements in comparison with existing systems are controlled (thanks to a set of checklists) and visualised via a radar diagram.

Part IV
The Roadmap and the State of the Art

Chapter 14
Evolution of Sustainability in Design Research and Practice

14.1 Introduction

In this last part we shall catch up with topics from previous chapters (but not only) in the light of the historical course of evolution of sustainability in design. First, we should outline the main stages of growing human awareness about environmental issues and sustainable development.

Historically, since the environmental issue was raised during the second half of the last century, the approach of mankind has moved from the end-of-pipe approach to actions increasingly aimed at prevention. In other words, we have moved from action and research focused exclusively on de-pollution systems, to research and innovation efforts that aim to reduce the cause of pollution at source (or more generally, of the environmental impact). The watchword of the United Nations Environmental Programme and other institutions became *cleaner production*, defined as "the continual redesigning of industrial processes and products to prevent pollution and the generation of waste, and risk for mankind and the environment".

The environmental issue, understood as the impact of the production–consumption system on the ecological equilibrium, started to be raised in the second half of the 1960s, as a consequence of the accelerating and spreading industrialisation. From those years we can recall the pollution of the Great Lakes of North America, the first ecological disasters caused by the washing of the cargo tanks of the oil tankers into the sea, or victims of smog in some industrial cities.

The first scientific works handling these problems were published at the beginning of the 1970s. International studies and debates started to consider the deterioration and exhaustion of natural resources to be an undesirable effect of industrial development. The natural limits of our planet became more clear, first in the light of uncontrollable technological productive development and then in the face of the increase in the world's population.

International debate about environmental issues intensified and spread further during the 1980s. The pressure from public opinion intensified and the institutions took their stand with a series of ecological norms and policies that examined the productive activities and were based on the *Polluter Pays Principle*.

An important study was drafted in 1987 by the UN World Commission on Environment and Development to give indications about the future of humanity. This report was called *Our Common Future* [1] and was the first to define sustainable development as "*a development that meets the needs of the present without compromising the ability of future generations to meet their own needs.*"

During the 1990s environmental matters entered the phase of maturity. *Caring for the Earth: A Strategy for Sustainable Living* – a publication for the World Conservation Union (IUCN) by the United Nations Environment Programme and the World Wide Fund for Nature (WWF) – has a competing definition of sustainable development: "*improving the quality of human life within the limits of capacity to protect the ecosystems.*" Thus, it accentuates the possibility of actually improving human life conditions while safeguarding the Earth's capacity to regenerate its resources.

These two definitions, considered together, describe the sustainable development as a practice that brings the benefits to human beings and ecosystems at the same time.

Another historic event of these years was the *United Nations Conference on Environment and Development (UNCED)* held in Rio de Janeiro in 1992. Since 1994, the sustainable development and environmental sustainability have formed a fundamental benchmark in the *5th Environmental Action Programme* of the European Commission.

This and other initiatives have provided the persistent integration of the concept of sustainable development with the documents of all international organisations, as a model for the reorientation of social and productive development.

In the second half of the 1990s, a series of studies and analyses led to a clearer understanding of the dimension of change necessary to achieve a society that is effectively and globally sustainable. Then, it was realised that conditions for sustainability can only be achieved by drastically reducing the consumption of environmental resources compared with the average consumption by advanced industrialised societies.

Some studies – taking into account demographic growth forecasts and assuming, rightly, an increase in the demand for well-being in currently disadvantaged countries – have shown a staggering result: in 50 years the conditions for sustainability will be achievable only by increasing the eco-efficiency of the production–consumption system by at least ten times.

It appears that there is a clear need for a profound transformation of our development concept, a system discontinuity. This scientific evidence leads to political, scientific, philosophical and social debate to question not only production processes, but artefacts in general, their design and development: products and services, systems and all the various forms of human settlement. Also, the patterns of

---

[1] *Our Common Future,* World Commission for Environment and Development (1987).

consumption and access to goods and services have to be discussed from the perspective of structural innovation[2].

Onwards from the 2000s (following the Johannesburg Conference and 10 years after Rio de Janeiro[3]) the necessity of the awareness and active engagement of all social participants involved in the production–consumption circuit is ever more present and pronounced.

Here, it accrues that a new generation of professionals with an interdisciplinary profile has to obtain the necessary knowledge and tools to know how to operate. The establishment of the *Decade of Education for Sustainable Development* (2005–2014) by the UN is a sign that the issue of the general updating of competences and integration of environmental prerequisites has found its way into international awareness.

From this highlighted course it has transpired that the potential role for design is growing, for the reason of extended interests in product and service innovation (not just the process any more) and even more because of the radical change in how the quality of a determined offer – its aesthetic and social dimension – is perceived. This increasing role has visibly and effectively entered in into design practice and research.

14.2 Evolution of Sustainability in Design

Let us try to shed some light on the different dimensions in which the design world has interpreted and developed sustainability.

The first level on which numerous theorists and academics have been working is the *selection of resources with a low environmental impact;* on the one hand the materials and on the other the energy sources. The main topics have been the elimination of toxic substances, recyclability, biodegradability and renewable resources.

Since the second half of the 1990s, attention has partially moved to the product level, to *the design of products with a low environmental impact*. In those years, it became clear which environmental effects are attributable to a product and how to assess them; the concept of *Life Cycle* was introduced and the concept of *functional unit* was re-contextualised in environmental terms.

Over the last few years, starting with a more stringent interpretation of environmental sustainability (which requires radical changes in production and consumption patterns), part of the debate has been reset to *design for sustainability starting from system innovation*, to a considerably wider dimension in respect of single products.

---

[2] *Cf.* UNEP, *Achieving Sustainable Consumption Patterns: the Role of the Industry* (1993); ERL, *Every Decision Counts* (1994); ERL, *The Best of Both Worlds* (1994).

[3] Both under the UN aegis.

Even more recently, design research opened a discussion on a possible role for design in social equity and in its various issues starting with the equity principle of a fair distribution and availability of natural resources.

Finally, and cross-wise with regard to the directions discussed so far, design for sustainability has started to *specialise* within certain sectors of commodity economics and in production–consumption contexts.

These directions will be discussed in more depth in the following paragraphs.

14.3 Low Impact Resources Selection

Though not traceable chronologically, the basic issues and criteria of low impact resources can be identified as they have entered into scientific and cultural debate, and later into design practices. With hindsight it is clear that this has taken place with an incorrect perception about their range and with few significant erroneous evaluations and interpretations (also design-wise).

One of the first issues is the *toxicity* and *harmfulness* of materials. These have already resulted in many regulations, but new updated estimates are still being accrued. In addition to traditional competence in actual design, it demands from the designer extended knowledge about correlated normatives and the actual adoption of a rather general *precautionary principle*.

Another associated question that has arisen with comparative ambiguity is the *naturalness* of the materials. This ambiguity, with its roots in terminology, which has been and still is accepted by many, claims that a "natural" material has by default no environmental impact whatsoever, or at least has a smaller one than synthetic materials. This argument, as it is understood now, is wrong for two reasons. First, in nature toxic and harmful substances are in abundance (even now nature is the cause of more toxic substances than humans, since they alter them only by re-contextualising them inside the mechanics of production and consumption). Second, practically all *natural* materials are subjected to a series of processes in order to become usable by production, and all those processes have their own environmental impact. Nowadays, however, the advantages of so-called natural materials are clear (even if the awareness in circles of practical planning is somewhat scarce): they are more renewable and generally more biodegradable than synthetic ones.

Another closely related topic that still has great influence among other great environmental problems is waste management, especially *recycling* materials and *incineration* in order to recover contained energy. Treating *design for recycling* properly demands transition from estimating the recyclability of materials to the economic and technological feasibility of the whole encompassed process. Thus, the design choices have to focus on the morphology and architecture of the product and design correlated to the entire path of the recyclable material. Therefore, design for recycling has to cover a set of indications that aim to facilitate every single stage: collection, transportation, disassembly and eventually cleaning, identification and production of secondary raw materials.

In juxtaposition with extending materials' lifespan emerges another concept: together with "know how-to-do" it is also important to "know how-to-undo". A significant event for this concept was the convention "Fare e disfare" ("To do and to undo") promoted by the Politecnico di Milano in 1993.

The reductive and insufficient effect of proclaiming a material recyclable becomes clear without designing the easy separation of incompatible materials or their easy recovery from discarded products, though even today it is common in design and communication circles to insist on the recyclability of materials, without taking into account any of this.

Another debate has been opened on the subject of *biodegradability*: an environmental quality that, similar to naturalness, has raised many misinterpretations. In fact, however important it is for the materials to be re-integrable with the ecosystems, for many products biodegradable materials might pose a problem in the sense of a premature expiration date; this in turn creates new productive and distributional processes for reasons of both substitution and discarding.

Last but not least comes the subject of *renewable resources* (energy-related and material) and research and development on different alternative sources like solar, wind, water, hydrogen and biomass power and their integration into (power-consuming) product systems.

Also, it has taken some time for this topic to be understood properly (not the actual design, but theoretically). This only happened when renewability has been associated with both the speed of recovery of the resource and the frequency of utilisation, more precisely, when it is understood that a resource is renewable only when it is replenished by natural processes at a rate comparable to its rate of consumption by humans.

14.4 Product Life Cycle Design

During the second half of the 1990s a new discipline in *the design of products with low environmental impact* emerged that was more concretely and realistically able to deal with the complexity of its subject: it became clear what is meant by the environmental requirements of industrial products; the concept of the *Life Cycle* was introduced and the concept of the *functional unit* was re-contextualised in environmental terms.

During these years, as a consequence of certain studies and new assessment methods, it became possible to evaluate the environmental impact of the input and output between the technosphere and the geosphere and the biosphere. This allowed us to specify what is meant by *the environmental requirements of industrial products*.

Among others, the methodology of *Life Cycle Assessment* (LCA) was established, which evaluates the environmental impact of the *input* and *output* of all processes at all life cycle stages that relate to the performance of the products, or what is called the *functional unit*.

Life Cycle Assessment was not born in the design circles; for this reason it also has its limits as a tool for designers. Nevertheless, it has strong influences on design research, which moreover has started to use the expression *Life Cycle Design* (LCD)[4], which actually deals with designing product life cycles. This expression is closely related to *ecodesign* and *design for environment*. Apart from the terminological issue, a new approach to product design was elaborated that entailed a larger vision than was traditionally used; it insisted on considering the whole life cycle within the design process, along with all processes required to produce the materials and then the product, to distribute, to use and finally to dispose of it, as a single unit. It also entails change in the design scale from single products to entire product systems, which means the sum of events that determine and accompany it during its life cycle.

It is a design that takes a systemic approach, allowing all the side-effects of a proposed product to be distinguished, including the stages traditionally not examined by design practice. A systematic idea of a product that reduces quantitatively and qualitatively the *input* of resources and energy, not to mention the impact of all emissions and waste, *i. e.* it calculates the harm of all impacts.

The second fundamental criterion of *Life Cycle Design* is designing the *function* that product has to supply rather than the product. Because of this association the overall analysis assesses whether the environmental impact has been minimised or reduced and by how much. Here, function, the fundamental topic for design culture (once a guideline, but criticised by many), gains a new meaning and vitality confronted with environmental issues[5].

Nowadays, the *environmental requirements for industrial design* or *Life Cycle Design* or *ecodesign* of the product have formed a structured discipline, and been provided with an established and definite theory, clear design guidelines[6], methods and tools, not to mention study courses in universities[7].

14.5 System Design for Eco-efficiency

As has been said before, over the last few years some design research centres, starting with a more stringent interpretation of environmental sustainability (that requires a systemic discontinuity in production and consumption patterns) have reset part of the debate on design for sustainability starting from system innovation.

[4] *Cf.* Keoleian and Menerey (1993).

[5] Tomas Maldonado speaks of "environmental neo-functionalism", 1998.

[6] It is generally accepted (irrespective of subtleties and importance in reports by different authors) that the environmental requirements of a product entail the following criteria: designing to reduce the amount of materials and energy necessary at any stage of the life cycle; choose atoxic and not harmful, materials, processes and energy sources with greatest renewability and the smallest exhaustibility as possible; to design and esteem recycling, incineration or composting of the discarded materials; designing for disassembly/separation of the materials and components.

[7] *Cf.* www.polimi.it/rapirete section of tools and section of courses; Costa and Vezzoli (2001).

Some authors have observed that the criteria for *product Life Cycle Design* meets obstacles in traditional supply models of product sales[8]; therefore, it is necessary to broaden the criteria to increase the probability of product innovation.

This new approach to system innovation has taken two different turns:

- Consideration of the systems as an integrated mix of products and services that are jointly capable of satisfying a given demand for well-being (systems in which attention is focused on the eco-efficiency of possible partnerships between socio-economic actors that take part in the value chain)[9].
- Consideration of the systems as open artificial ecosystems trying to minimise the waste and emissions (systems of industrial symbiosis in which the attention is focused on the resource flows of input and output of different types of products)[10].

As a matter of fact, such a definition of system innovation remains within the fundamental criteria of *Life Cycle Design*, though it brings out more honestly (as a principle matter) the need for reconfiguration of the system in order to get the results. Environmental value is still estimated as a summarised impact of all life cycles of the products and services that constitute the supply system, as well as their functional unity.

Concisely, the fragmentation of the social actors along the life cycle of the product (in a traditional economy of industrial societies) excludes the eco-efficiency of the life cycle system coinciding with the economic interests of participants.

This is true in the case of producers of a product service system answering the demand, as well as in the case of producers not connected to this relationship, but connected in terms of cascaded utilisation of waste and emissions of the production processes, in which case we are speaking about system eco-efficiency resulting from new convergences (with economic interest) between the stakeholders: innovations not any longer on the product level, but of new relationships between various *actors*.

So, it rather remains within the competence of strategic design, participatory design and the design of new forms of partnership between different stakeholders belonging to a particular *value constellation* [11] or symbiotically connected production processes [12].

As a matter of fact, these discourses have drawn together environmentally sustainable design and strategic design, and have also coined a new term: *strategic design for sustainability* [13]. Underlining the fact that the design for environmental sustainability has to make use of and integrate the methods and tools of strategic design (and *vice versa*).

---

[8] *Cf.* Stahel (2001).
[9] *Cf.* Bistagnino, *Design con futuro* (2003).
[10] *Cf.* UNEP, *Product-Service Systems and Sustainability. Opportunities for Sustainable Solutions* (2002).
[11] *Cf.* Normann and Ramirez (1995).
[12] *Cf.* Bistagnino and Lanzavecchia (2000).
[13] *Cf.* Manzini and Vezzoli (2001).

14.6 Design for Social Equity and Cohesion

The debate about more sustainable consumption patterns has been included in the agenda of the major international governmental institutions over recent years, starting with the United Nations. Particularly significant was the setting up of the *Sustainable Consumption Unit* of the UNEP (United Nation Environmental Programme) in May 2000. The initial assumption was that "in spite of the progress made by the industrial world and companies during the last decade [...] the extent to which consumption exceeds the Earth's capacity to supply resources and absorb waste and emissions is still dramatically evident"[14]. This analysis also recognises (exactly as in the conceptual assumptions of sustainable development) the so-called equity principle, which proposes that every person, in a fair distribution of resources, has a right to the same *environmental space*, *i. e.* to the same availability of global natural resources (or better, to the same level of satisfaction that can be derived from them in one way or another).

These analyses belong to social and ethical territory.

The European Union has defined a concept of *social equity and cohesion* within the *Sustainable Development Strategies* [15].

This complex issue can be summarised in the following question: how can we foster new quality criteria so as to separate the social demand for well-being from a relationship that is directly proportional to the increase in consumption of resources, characteristic of mature industrialised societies? And how can this process fit into and orientate transitions that are already in progress in mature industrialised societies, and those that are desirable for developing newly industrialised countries?

From the design point of view it is important to identify which answers to these questions could be given by design culture.

The few theoretical contributions made by design culture in the field of consumption are not necessarily recent. By the end of the 1960s, for various reasons, the theory and culture of design in Italy anticipated a critique of consumption patterns[16], or at least some of the leading figures in the culture of design acted as spokespersons for issues relating to the responsibility of designers for consumption patterns, although in different ways and not directly and exclusively associated with environmental impact. We can recall the criticism of consumer society made in denouncements by some Italian exponents of *Radical Design* on the one hand, and the reaction of Tomas Maldonado on the other, who appealed to a new "design hope". The question of designer responsibility was again brought up at the beginning of the 1970s, though was never resolved with regard to its implications for design practice. Victor Papanek and Tomas Maldonado express similar positions, as far as the role of consumption is concerned. Papanek writes: "design can and must become a means for young people to take part in the transformation of society[17]".

---

[14] *Cf.* Geyer-Allely (2002).
[15] *Cf.* EU (2006).
[16] For further study of the Italian situation, *cf.* Pietroni and Vezzoli (2004).
[17] *Cf.* Papanek (1973).

Recently, the debate has re-started about design's place in the socio-ethical dimension of sustainability, whose limits and implications are still not exactly traceable. Still, we can claim that when the issue of sustainable consumption crosses that of socio-ethical sustainability, the spectrum of implications, of responsibilities, extends to several different issues such as[18]:

- The principles and rules of democracy, human rights and freedom
- The achievement of peace and security
- The reduction of poverty and injustice
- Improved access to information, training and employment
- Respect for cultural diversity, regional identity and natural biodiversity

This is an extremely vast and complex issue and its implications for design have so far been analysed very little (and are difficult to face without falling into easy, barely constructive moralism). We can observe new, though sporadic, interest on the part of design research to move into this territory.

One example of a working hypothesis (created by an international group of experts and promoted by UNEP) is to identify the possibilities of extending and applying the concept of a eco-efficient product Service system in developing countries[19]. To facilitate the process of socio-economic development of an emerging context – by jumping over or by-passing the stage characterised by individual consumption/ownership of mass produced goods (as in industrialised countries) – towards a more advanced service economy, "satisfaction-based" and low resources-intensive (which is starting to take its place in industrial countries).

Another related example is the hypothesis that design could amplify the phenomenon of responsible economy (even in advanced industrialised contexts). Research has increasing interests in various forms of *creative communities*[20], characterised by the self-organised activities of aware, critical, motivated citizens who are organised to a greater or lesser extent into networks and districts of economic solidarity. It is linked to sustainable social innovation, *i. e.* solutions of high social quality and low environmental impact that spring from active, bottom-up, social participation, especially that concerning exploration of different interactions, integrations and synergies between real cases in urban space and professionals from sustainable design and development[21].

The common element of these two areas seems to be the *strategic design* approach, or even the design of innovative (and sustainable) relationships between socio-economic actors.

From these first modest and limited trials it can be foreseen that design for socio-ethical sustainability represents an area (very complex, but also stimulatory and rewarding), where design research has to contribute with working hypotheses and disciplinary responses.

---

[18] *Cf. World Summit on Sustainable Development. Draft Political Declaration*, 2002.
[19] *Cf.* UNEP, 2002.
[20] *Cf.* Florida 2002; Manzini Manzini, Jegou, 2003.
[21] *Cf.* Vezzoli, 2003.

14.7 State of the Art

When these design directions are assumed to interpret the discipline's history, it appears a well-articulated panorama. Therefore, it is important to demonstrate what points they have reached in design research and in design education, and where they could (and should) proceed. For this purpose it can be useful to map these directions respectively measuring the *disciplinary consolidation* (drawn from the results of research) and the *diffusion* of related study possibilities (offered by design universities). When looking at the diagram in Fig. 14.1, the new fronts of research can be found in the bottom left-hand corner (towards 0% research and teaching), and in the upper right-hand corner (towards 100% research and teaching) should be the various existing fields of the discipline, *i. e.* they should be well studied and scrupulously taught.

In this chart, the *low impact material/energy selection and the LCD/eco-design of the product* can be seen as fairly well consolidated and discreetly penetrated within design practices.

For *system design for eco-efficiency*, the level of consolidation is inferior and practice is, logically, far more sporadic.

On the *design for social equity and cohesion* front little has been elaborated at a technical level (it is, as we say, a new research frontier) and there are obviously very few teaching proposals.

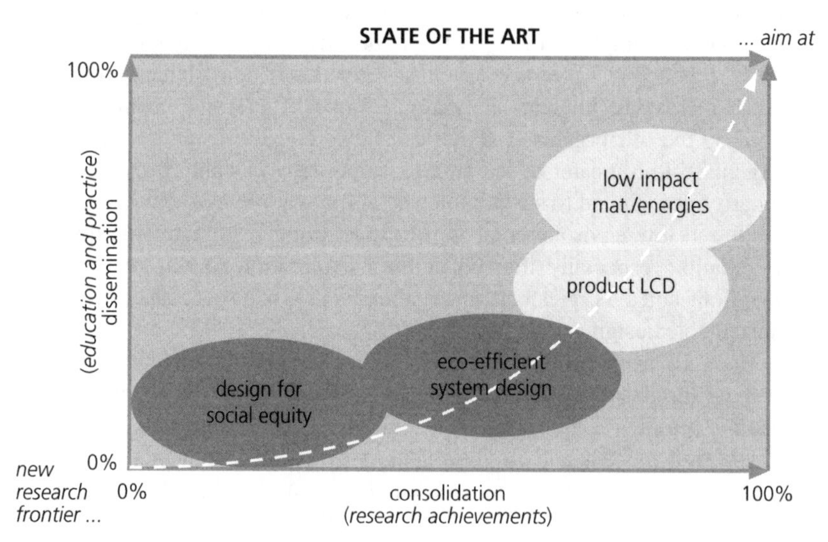

Fig. 14.1 Map of the *disciplinary consolidation* and the *diffusion* of related research achievements (Vezzoli, 2007)

Appendix A
Design Criteria and Guidelines

A.1 Minimise Materials Consumption

Minimise material content:

- *Dematerialise* the product or some of its components
- Digitalise the product or some of its components
- Miniaturise
- Avoid over-sized dimensions
- Reduce thickness
- Apply ribbed structures to increase structural stiffness
- Avoid extra components with little functionality

Minimise scraps and discards:

- Select processes that reduce scraps and discarded materials during production
- Engage simulation systems to optimise transformation processes

Minimise or avoid packaging:

- Avoid packaging
- Apply materials only where absolutely necessary
- Design the package to be part (or to become a part) of the product

Engage more consumption-efficient systems:

- Design for more efficient consumption of operational materials
- Design for more efficient supply of raw materials
- Design for more efficient use of maintenance materials
- Design systems for consumption of passive materials

- Design for cascading recycling systems
- Facilitate the user to reduce materials consumption
- Set the product's default state at minimal materials consumption

Engage systems of flexible materials consumption:

- Engage digital support systems with dynamic configuration
- Design dynamic materials consumption for different operational stages
- Engage sensors to adjust materials consumption according to differentiated operational stages
- Reduce resource consumption in the product's default state

Minimise materials consumption during the product development phase:

- Minimise the consumption of stationery goods and their packages
- Engage digital tools in designing, modelling and prototype creation
- Engage digital tools for documentation, communication and presentation

A.2 Minimising Energy Consumption

Minimise energy consumption during pre-production and production:

- Select materials with low energy intensity
- Select processing technologies with the lowest energy consumption possible
- Engage efficient machinery
- Use heat emitted in processes for preheating other determined process flows
- Engage pump and motor speed regulators with dynamic configuration
- Equip the machinery with intelligent power-off utilities
- Optimise the overall dimensions of the engines
- Facilitate engine maintenance
- Define accurately the tolerance parameters
- Optimise the volumes of required real estate
- Optimise stocktaking systems
- Optimise transportation systems and scale down the weight and dimensions of all transportable materials and semi-products
- Engage efficient general heating, illumination and ventilation in buildings

Minimise energy consumption during transportation and storage:

- Design compact products with high storage density
- Design concentrated products
- Equip products with onsite assembly
- Scale down the product weight
- Scale down the packaging weight

- Decentralise activities to reduce transportation volumes
- Select local material and energy sources

Select systems with energy-efficient operation stage:

- Design attractive products for collective use
- Design for energy-efficient operational stages
- Design for energy-efficient maintenance
- Design systems for consumption of passive energy sources
- Engage highly efficient energy conversion systems
- Design/engage highly efficient engines
- Design/engage highly efficient power transmission
- Use highly caulked materials and technical components
- Design for localised energy supply
- Scale down the weight of transportable goods
- Design energy recovery systems
- Design energy-saving systems

Engage dynamic consumption of energy:

- Engage digital dynamic support systems
- Design dynamic energy consumption systems for differentiated operational stages
- Engage sensors to adjust consumption during differentiated operational stages
- Equip machinery with intelligent power-off utilities
- Program product's default state at minimal energy consumption

Minimise energy consumption during product development:

- Engage efficient workplace heating, illumination and ventilation
- Engage digital tools for communicating with remote working sites

A.3 Minimising Toxic Emissions

Select non-toxic and harmless materials:

- Avoid toxic or harmful materials for product components
- Minimise the hazard of toxic and harmful materials
- Avoid materials that emit toxic or harmful substances during pre-production
- Avoid additives that emit toxic or harmful substances
- Avoid technologies that process toxic and harmful materials
- Avoid toxic or harmful surface treatments
- Design products that do not consume toxic and harmful materials
- Avoid materials that emit toxic or harmful substances during usage
- Avoid materials that emit toxic or harmful substances during disposal

Select non toxic and harmless energy resources:

- Select energy resources that reduce dangerous emissions during pre-production and production
- Select energy resources that reduce dangerous emissions during distribution
- Select energy resources that reduce dangerous emissions during usage
- Select energy resources that reduce dangerous residues and toxic and harmful waste

A.4 Renewable and Bio-compatible Resources

Select renewable and bio-compatible materials:

- Use renewable materials
- Avoid exhaustive materials
- Use residual materials of production processes
- Use retrieved components from disposed products
- Use recycled materials, alone or combined with primary materials
- Use bio-degradable materials

Select renewable and bio-compatible energy resources:

- Use renewable energy resources
- Engage the cascade approach
- Select energy resources with high second-order efficiency

A.5 Optimisation of Product Lifespan

Design appropriate lifespan:

- Design components with co-extensive lifespan
- Design lifespan of replaceable components according to scheduled duration
- Select durable materials according to the product performance and lifespan
- Avoid selecting durable materials for temporary products or components

Reliability design:

- Reduce overall number of components
- Simplify products
- Eliminate weak *liaisons*

Facilitate upgrading and adaptability:

- Enable and facilitate software upgrading
- Enable and facilitate hardware upgrading

- Design modular and dynamically configured products to facilitate their adaptability for changing environments
- Design multifunctional and dynamically configured products to facilitate their adaptability for changing cultural and physical individual backgrounds
- Design onsite upgradeable and adaptable products
- Design complementary tools and documentation for product upgrading and adaptation

Facilitate maintenance:

- Simplify access and disassembly to components to be maintained
- Avoid narrow slits and holes to facilitate access for cleaning
- Prearrange and facilitate the substitution of short-lived components
- Equip the product with easily usable tools for maintenance
- Equip products with diagnostic and/or auto-diagnostic systems for maintainable components
- Design products for easy on-site maintenance
- Design complementary maintenance tools and documentation
- Design products that need less maintenance

Facilitate repairs:

- Arrange and facilitate disassembly and re-attachment of easily damageable components
- Design components according to standards to facilitate substitution of damaged parts
- Equip products with automatic damage diagnostics system
- Design products for facilitated onsite repair
- Design complementary repair tools, materials and documentation

Facilitate re-use:

- Increase the resistance of easily damaged and expendable components
- Arrange and facilitate access and removal of retrievable components
- Design modular and replaceable components
- Design components according to standards to facilitate replacement
- Design re-usable auxiliary parts
- Design the re-filling and re-usable packaging
- Design products for secondary use

Facilitate re-manufacture:

- Design and facilitate removal and substitution of easily expendable components
- Design structural parts that can be easily separated from external/visible ones
- Provide easier access to components to be re-manufactured
- Calculate accurate tolerance parameters for easily expendable connections
- Design for excessive use of materials in places more subject to deterioration
- Design for excessive use of material for easily deteriorating surfaces

A.6 Improve Lifespan of Materials

Adopt the cascade approach:

- Arrange and facilitate recycling of materials in components with lower mechanical requirements
- Arrange and facilitate recycling of materials in components with lower aesthetical requirements
- Arrange and facilitate energy recovery from materials throughout combustion

Select materials with most efficient recycling technologies:

- Select materials that easily recover after recycling the original performance characteristics
- Avoid composite materials or, when necessary, choose easily recyclable ones
- Engage geometrical solutions like ribbing to increase polymer stiffness instead of reinforcing fibres
- Prefer thermoplastic polymers to thermosetting
- Prefer heat-proof thermoplastic polymers to fireproof additives
- Design considering the secondary use of the materials once recycled

Facilitate end-of-life collection and transportation:

- Design in compliance with product retrieval system
- Minimise overall weight
- Minimise cluttering and improve stackability of discarded products
- Design for the compressibility of discarded products
- Provide the user with information about the disposing modalities of the product or its parts

Material identification:

- Codify different materials to facilitate their identification
- Provide additional information about the material's age, number of times recycled in the past and additives used
- Indicate the existence of toxic or harmful materials
- Use standardised materials identification systems
- Arrange codifications in easily visible places
- Avoid codifying after component production stages

Minimise the number of different incompatible materials:

- Integrate functions to reduce the overall number of materials and components
- *Monomaterial* strategy: only one material per product or per sub-assembly
- Use only one material, but processed in sandwich structures
- Use compatible materials (that could be recycled together) within the product or sub-assembly
- For joining use the same or compatible materials as in components (to be joined)

Facilitate cleaning:

- Avoid unnecessary coating procedures
- Avoid irremovable coating materials
- Facilitate removal of coating materials
- Use coating procedures that comply with coated materials
- Avoid adhesives or choose ones that comply with materials to be recycled
- Prefer the dyeing of internal polymers, rather than surface painting
- Avoid using additional materials for marking or codification
- Mark and codify materials during moulding
- Codify polymers using lasers

Facilitate composting:

- Select materials that degrade in the expected end-of-life environment
- Avoid combining non-degradable materials with products that are going to be composted
- Facilitate the separation of non-degradable materials

Facilitate combustion:

- Select high energy materials for products that are going to be incinerated
- Avoid materials that emit dangerous substances during incineration
- Avoid additives that emit dangerous substances during incineration
- Facilitate the separation of materials that would compromise the efficiency of combustion (with low energy value)

A.7 Design for Disassembly

Reduce and facilitate operations of disassembly and separation:

Overall architecture:

- Prioritise the disassembly of toxic and dangerous components or materials
- Prioritise the disassembly of components or materials with higher economic value
- Prioritise the disassembly of more easily damageable components
- Engage modular structures
- Divide the product into easily separable and manipulatable sub-assemblies
- Minimise overall dimensions of the product
- Minimise hierarchically dependent connections between components
- Minimise different directions in the disassembly route of components and materials
- Increase the linearity of the disassembly route
- Engage a sandwich system of disassembly with central joining elements

Shape of components and parts:

- Avoid difficult-to-handle components
- Avoid asymmetrical components, unless required
- Design leaning surfaces and grabbing features in compliance with standards
- Arrange leaning surfaces around the product's centre of gravity
- Design for easy centring on the component base

Shape and accessibility of joints:

- Avoid joining systems that require simultaneous interventions for opening
- Minimise the overall number of fasteners
- Minimise the overall number of different fastener types (that demand different tools)
- Avoid difficult-to-handle fasteners
- Design accessible and recognisable entrances for dismantling
- Design accessible and controllable dismantling points

Engage reversible joining systems

- Employ two-way snap-fit
- Employ joints that are opened with common tools
- Employ joints that are opened with special tools, when opening could be dangerous
- Design joints made of materials that become reversible only in determined conditions
- Use screws with hexagonal heads
- Prefer removable nuts and clips to self-tapping screws
- Use screws made of materials compatible with joint components, to avoid their separation before recycling
- Use self-tapping screws for polymers to avoid using metallic inserts

Engage easily collapsible permanent joining systems:

- Avoid rivets on incompatible materials
- Avoid staples on incompatible materials
- Avoid additional materials while welding
- Weld with compatible materials
- Prefer ultrasonic and vibration welding with polymers
- Avoid gluing with adhesives
- Employ easily removable adhesives

Co-design special technologies and features for crushing separation:

- Design thin areas to enable the taking off of incompatible inserts, by pressurised demolition
- Co-design cutting or breaking paths with appropriate separation technologies for incompatible materials separation

- Equip the product with a device to separate incompatible materials
- Employ joining elements that allow their chemical or physical destruction
- Make the breaking points easily accessible and recognisable
- Provide the products with information for the user about the characteristics of crushing separation

  Use materials that are easily separable after being crushed.
  Use additional parts that are easily separable after crushing of materials.

Appendix B
Diagrams of Environmental Impacts

The following pages present diagrams of the following environmental effects.

1. Global warming
2. Ozone layer depletion
3. Acidification
4. Eutrophication
5. Summer smog
6. Winter smog
7. Toxic air pollution
8. Toxic water and ground pollution
9. Landfills

Every diagram presents the *agents* (emissions) and their *impact* on the geosphere and biosphere, and the detrimental *effects* that can occur. Likewise, the direct and indirect causes of these emissions are presented.

B.1 Global warming: greenhouse effect

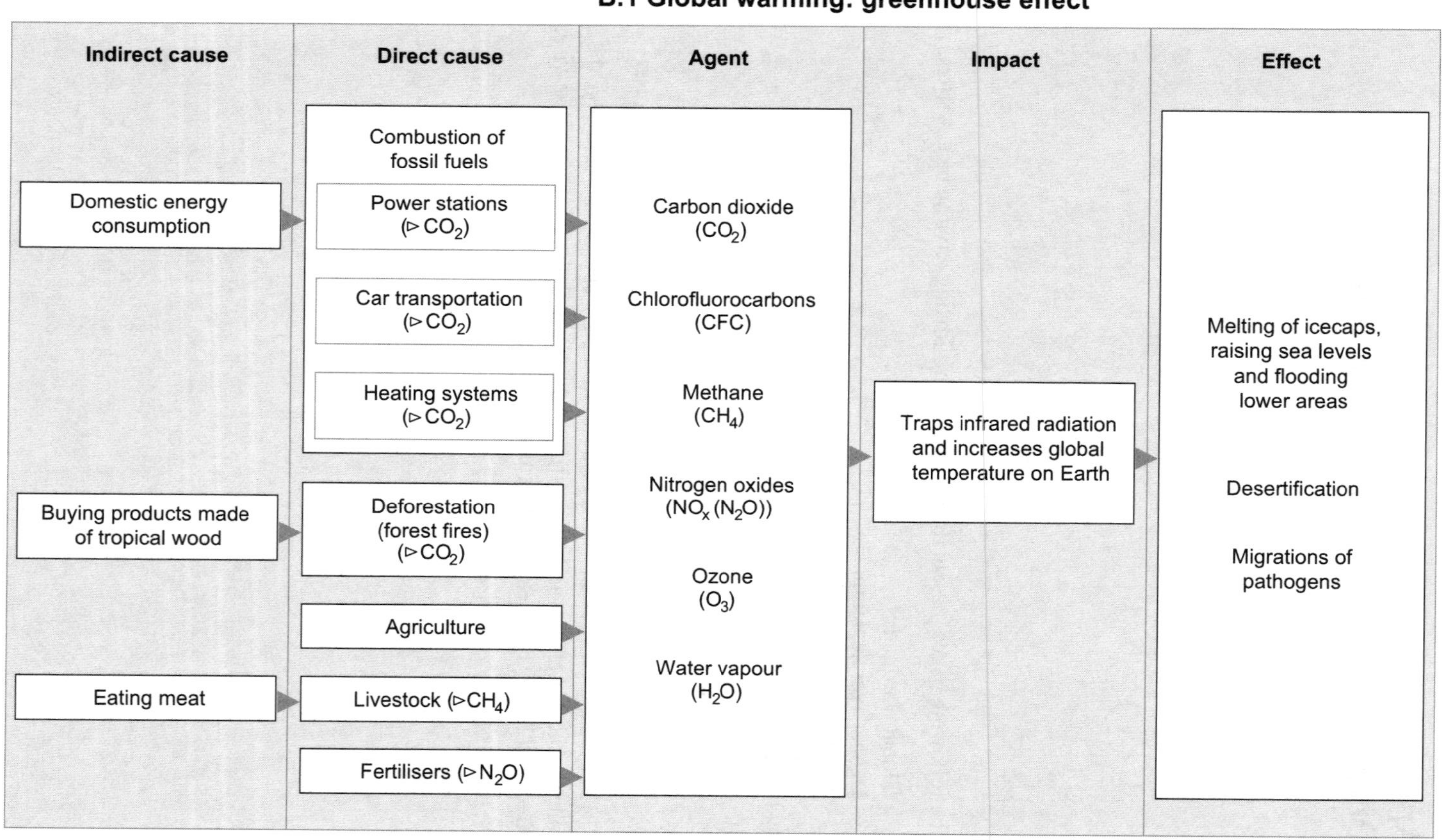

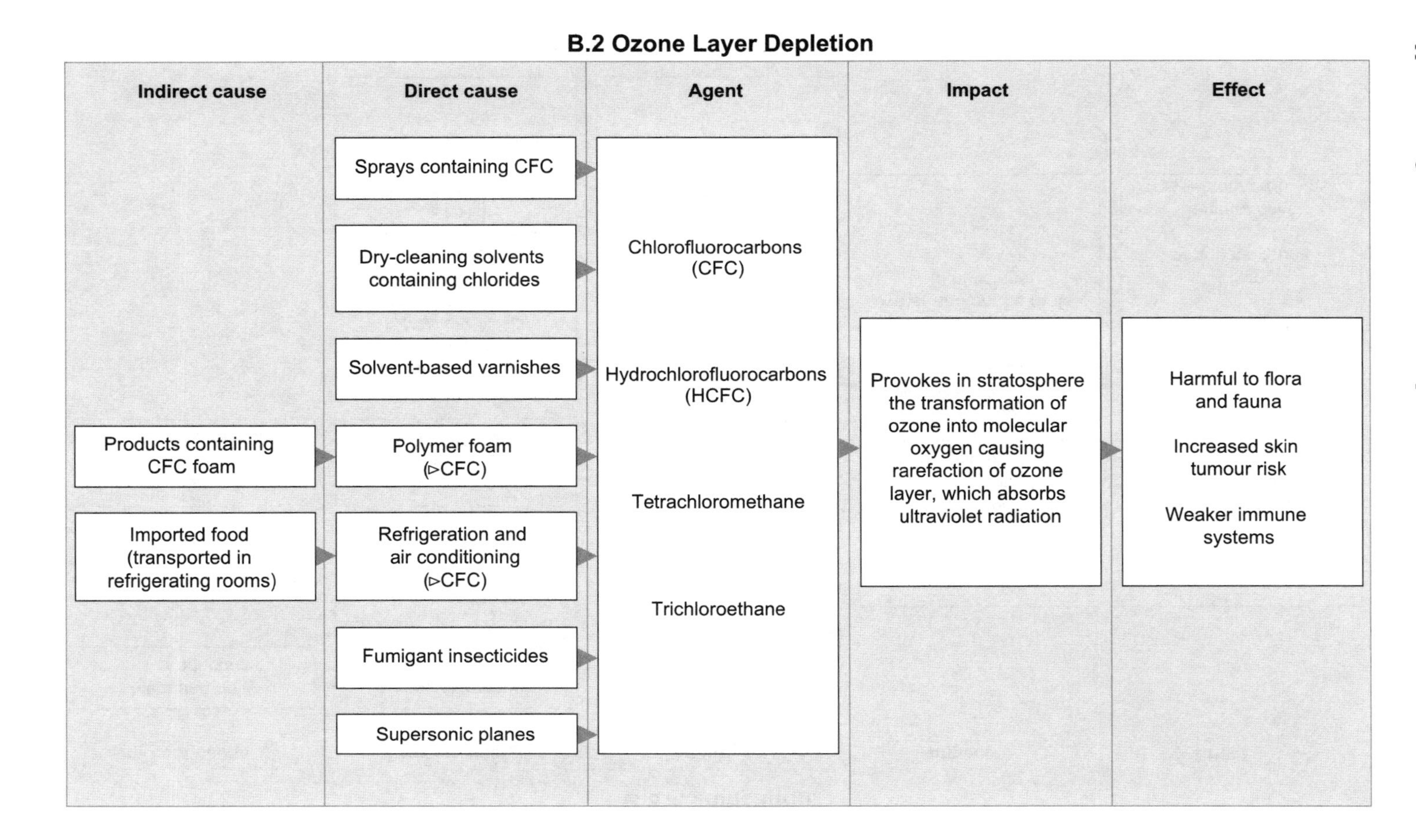
B.2 Ozone Layer Depletion
Indirect cause
Direct cause
Agent
Impact
Effect
Products containing CFC foam
Imported food (transported in refrigerating rooms)
Sprays containing CFC
Dry-cleaning solvents containing chlorides
Solvent-based varnishes
Polymer foam (⊳CFC)
Refrigeration and air conditioning (⊳CFC)
Fumigant insecticides
Supersonic planes
Chlorofluorocarbons (CFC)
Hydrochlorofluorocarbons (HCFC)
Tetrachloromethane
Trichloroethane
Provokes in stratosphere the transformation of ozone into molecular oxygen causing rarefaction of ozone layer, which absorbs ultraviolet radiation
Harmful to flora and fauna
Increased skin tumour risk
Weaker immune systems

B.3 Acidification

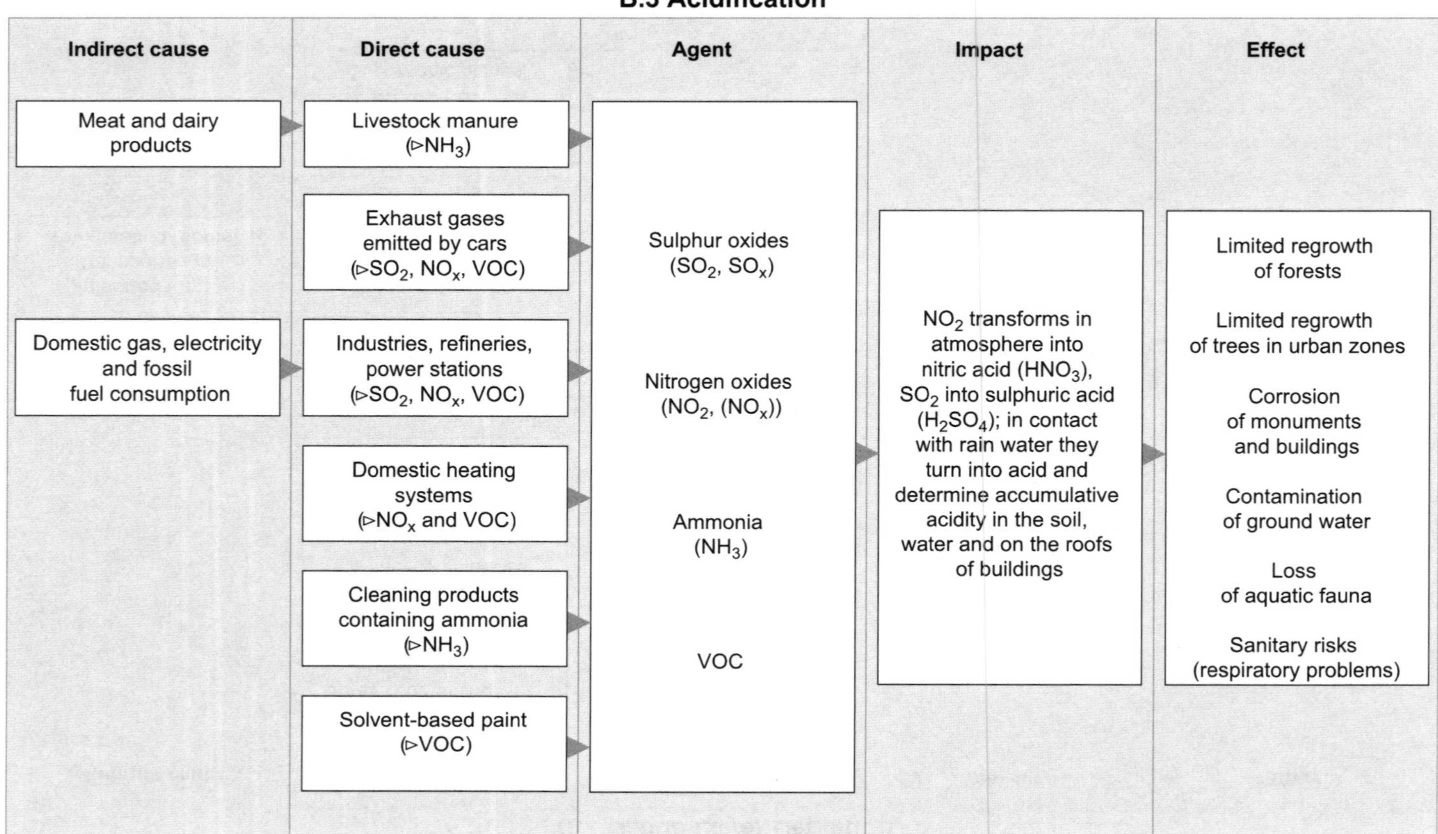

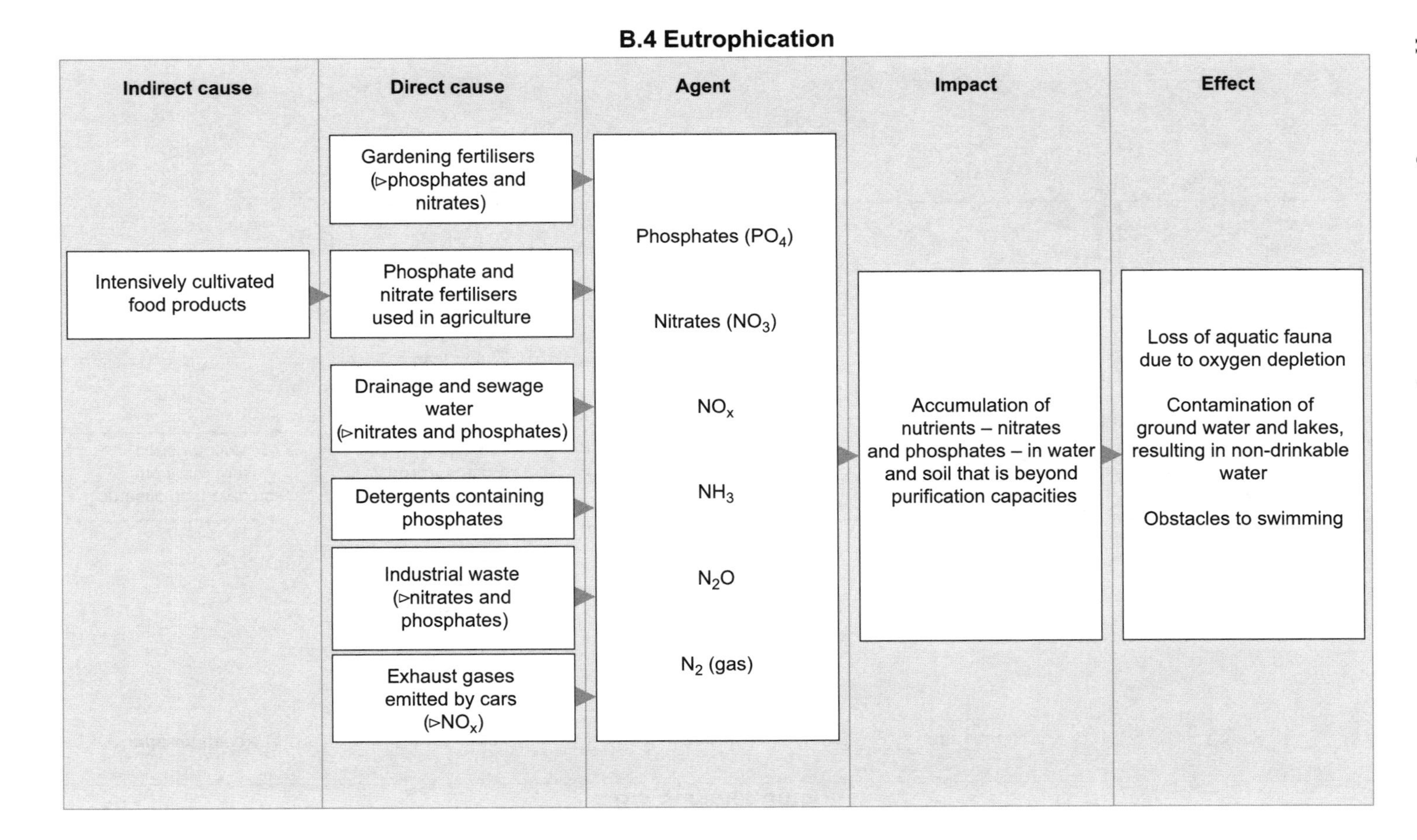
B.4 Eutrophication
Indirect cause
Direct cause
Agent
Impact
Effect
Intensively cultivated food products
Gardening fertilisers (▷phosphates and nitrates)
Phosphate and nitrate fertilisers used in agriculture
Drainage and sewage water (▷nitrates and phosphates)
Detergents containing phosphates
Industrial waste (▷nitrates and phosphates)
Exhaust gases emitted by cars (▷NO_x)
Phosphates (PO_4)
Nitrates (NO_3)
NO_x
NH_3
N_2O
N_2 (gas)
Accumulation of nutrients – nitrates and phosphates – in water and soil that is beyond purification capacities
Loss of aquatic fauna due to oxygen depletion
Contamination of ground water and lakes, resulting in non-drinkable water
Obstacles to swimming

B.5 Summer Smog

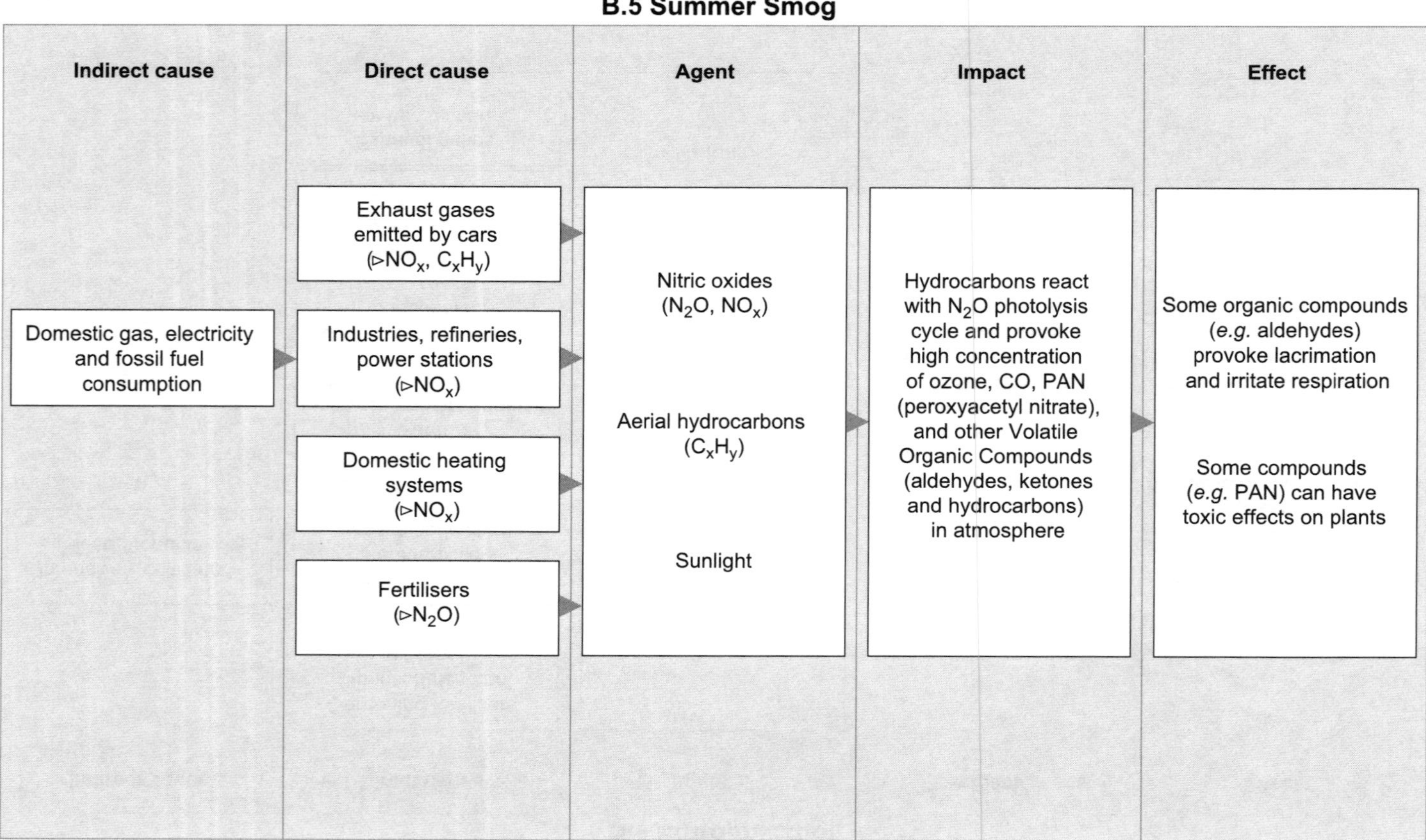

B.6 Winter Smog

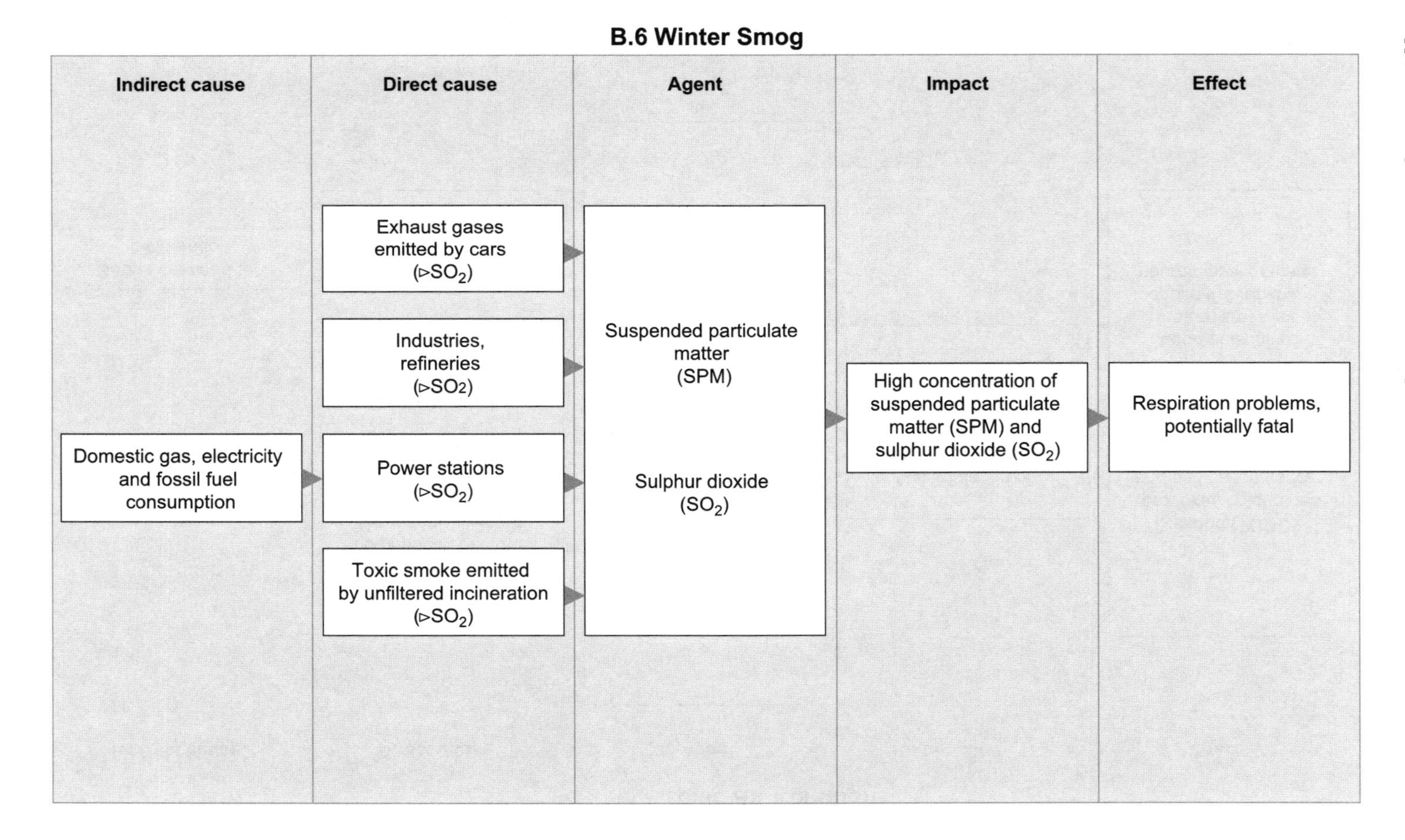

B.7 Toxic Air Pollution

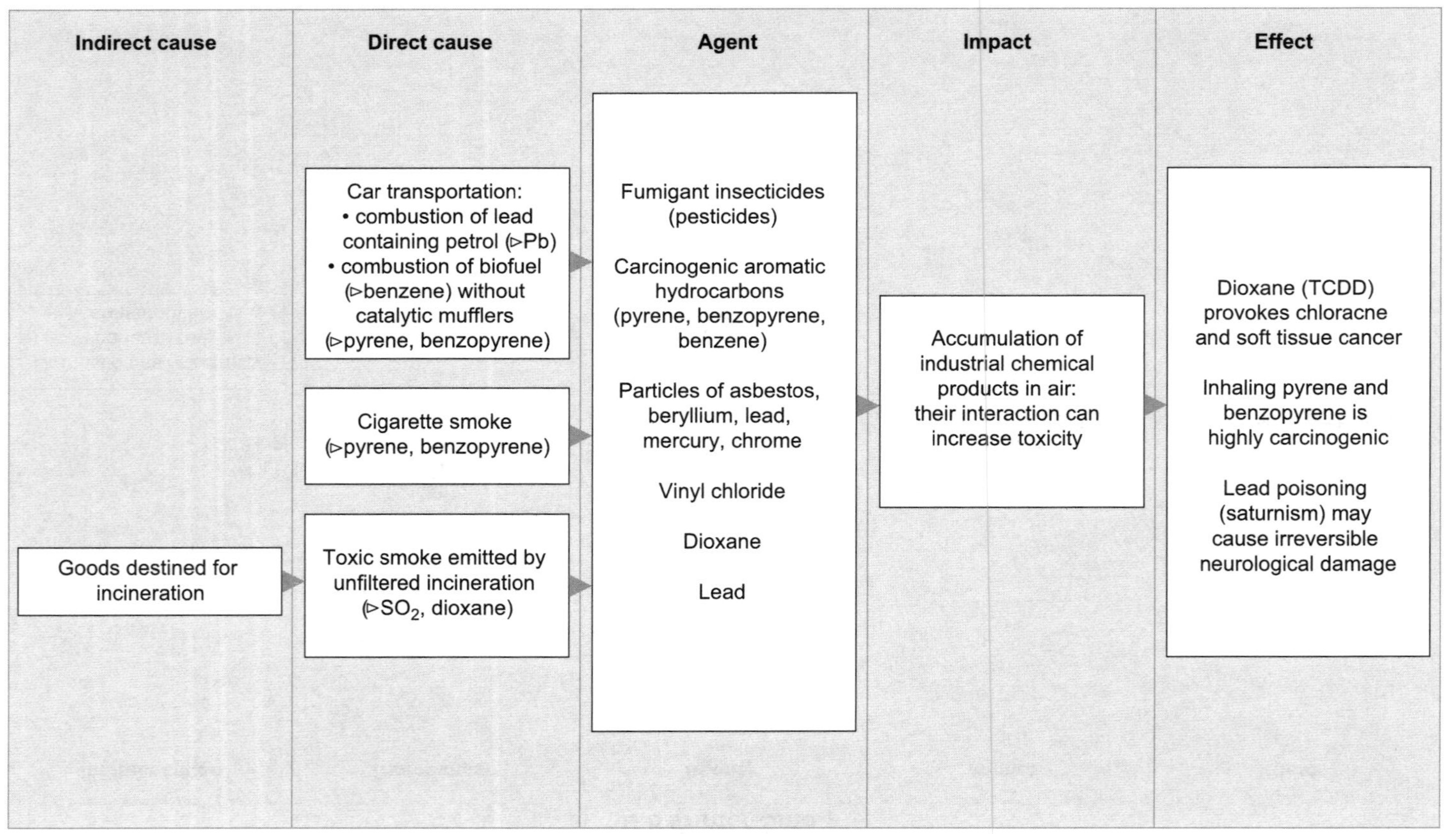

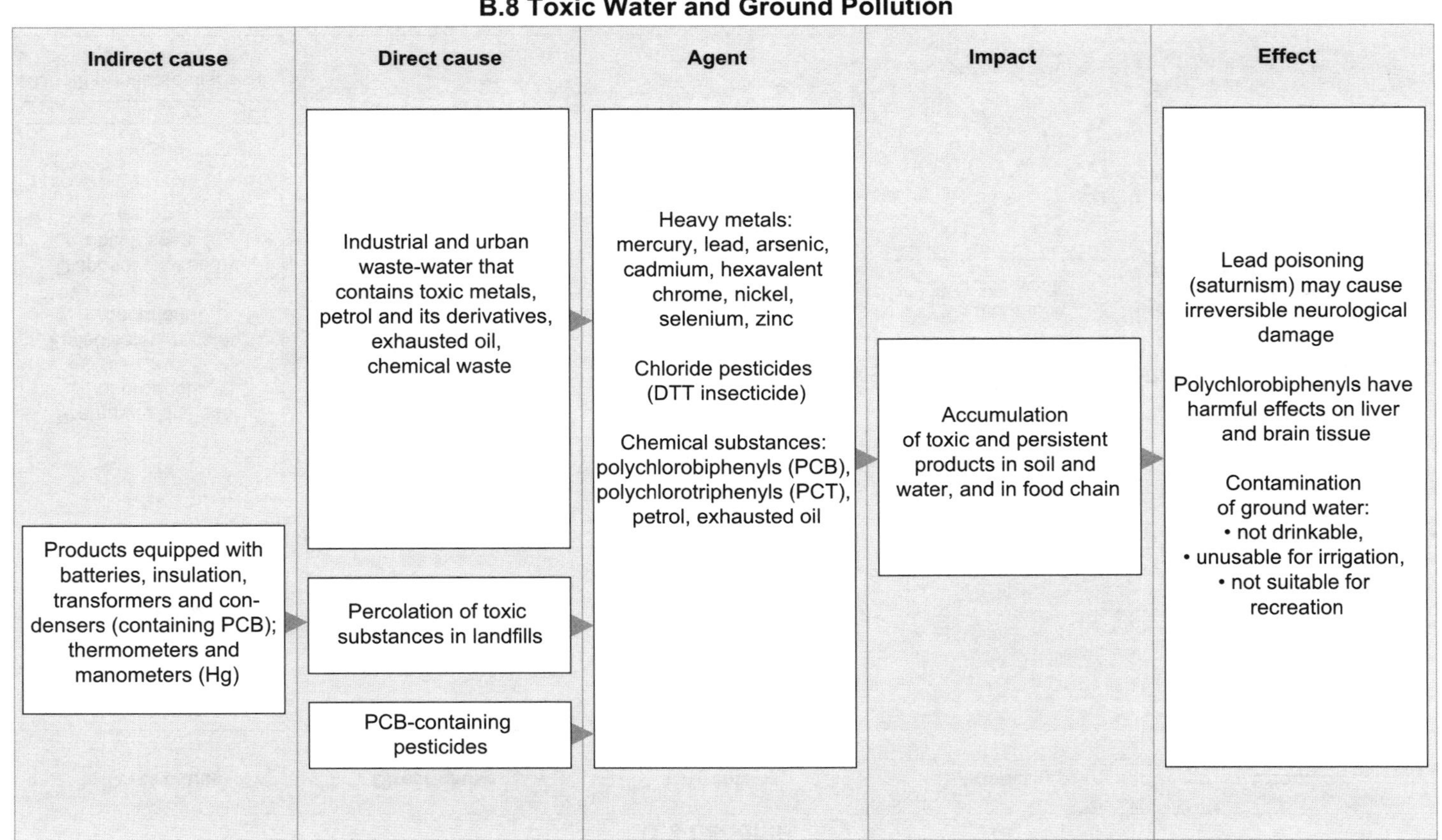
B.8 Toxic Water and Ground Pollution
Indirect cause
Direct cause
Agent
Impact
Effect
Products equipped with batteries, insulation, transformers and condensers (containing PCB); thermometers and manometers (Hg)
Industrial and urban waste-water that contains toxic metals, petrol and its derivatives, exhausted oil, chemical waste
Percolation of toxic substances in landfills
PCB-containing pesticides
Heavy metals: mercury, lead, arsenic, cadmium, hexavalent chrome, nickel, selenium, zinc
Chloride pesticides (DTT insecticide)
Chemical substances: polychlorobiphenyls (PCB), polychlorotriphenyls (PCT), petrol, exhausted oil
Accumulation of toxic and persistent products in soil and water, and in food chain
Lead poisoning (saturnism) may cause irreversible neurological damage
Polychlorobiphenyls have harmful effects on liver and brain tissue
Contamination of ground water:
• not drinkable,
• unusable for irrigation,
• not suitable for recreation

B.9 Landfills

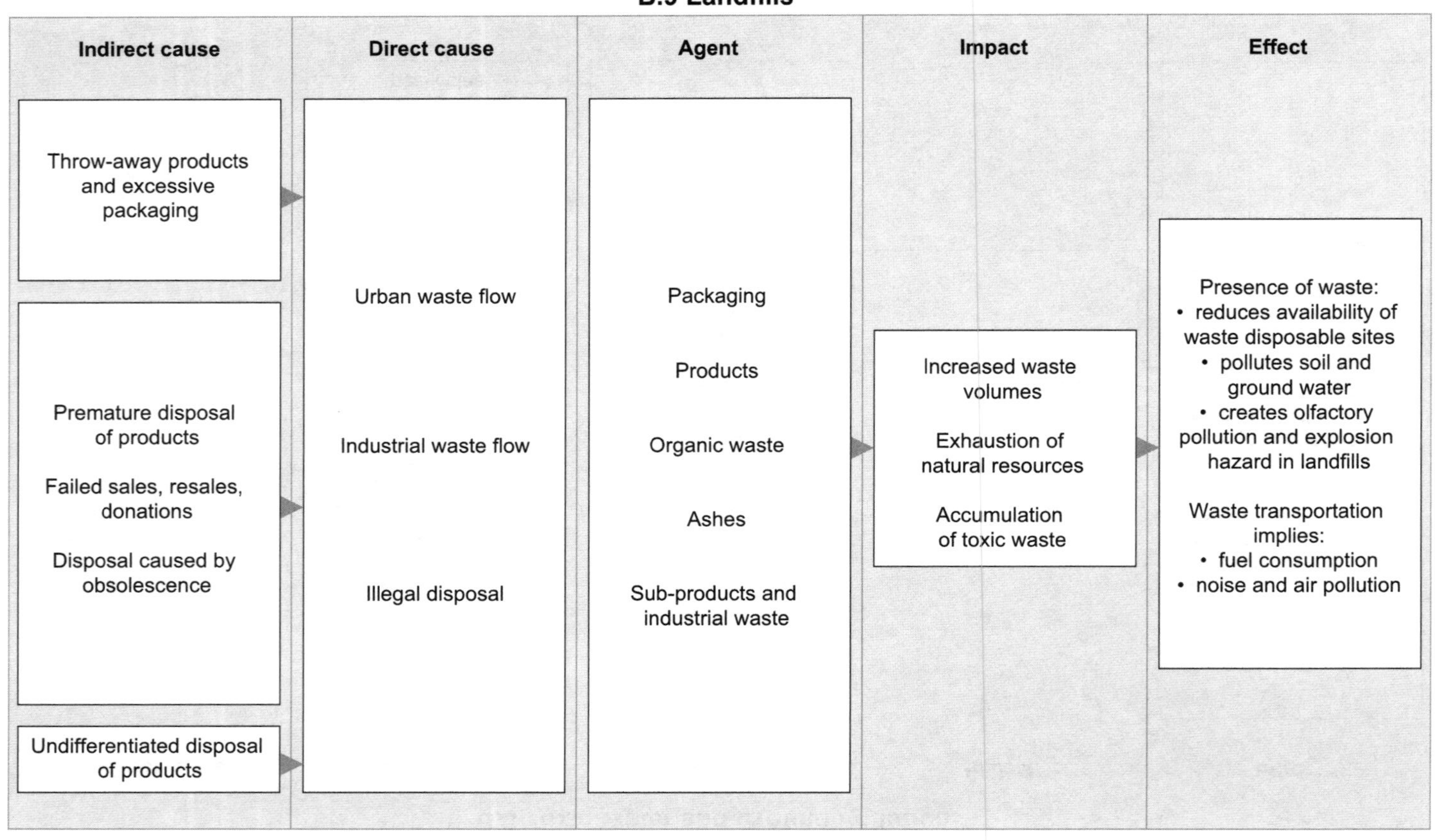

References

Part I

Augé M (1992) Les non-lieux. Seoul, Paris
Balbo L (ed) (1993) Friendly 93. Anabasi, Milan
Baumann Z (1998) Globalization. The human consequences. Polity Press, Cambridge/Blackwell, Oxford
Bauman Z (1999) La società dell'incertezza. Il Mulino, Bologna
Baumann Z (2000) Liquid modernity. Polity Press, Cambridge/Blackwell, Oxford
Beck U (1997) Was ist Globalisierung? Irrtümer des Globalisierung – Antworten auf Globalisierung. Suhrkamp, Frankfurt am Main
Benkler Y (2006) The wealth of networks. Yale University Press, New Haven
Bocchi G, Ceruti M (eds) (1985) La sfida della complessità. Feltrinelli, Milan
Braungart M, Englefried J (1992) An "Intelligent Product System" to replace "Waste Management". Fresenius Environ Bull 1:613–619
Braungart M, McDonough A (1998) The next industrial revolution. Atlantic Monthly, October
Brezet JC, Bijma AS, Ehrenfeld J, Silvester S (2001) The design of eco-efficient services. Design for sustainable program. Faculty OCP, Delft University of Technology, Delft
Brezet JC, van Hemel CG (1997) UNEP Eco-design manual. Ecodesign: a promising approach to sustainable production and consumption. UNEP, Paris
Bright C, Flavin C, Platt McGinn A, et al. (2003) State of the World 2003. A Worldwatch Institute Report on progress toward a sustainable society. Worldwatch Institute, Washington, DC
Brown L (2003) Eco-economy: building an economy for the earth. Earthscan, London
Capra F (1996) The web of life. HarperCollins, London
Capra F (2002) The hidden connections. A science for sustainable living. HarperCollins, London
Castells M (1996) The rise of the network society. The information age: economy, society and culture, vol. 1. Blackwell, Oxford
Ceruti M, Laszlo E (ed) (1988) Physis: abitare la terra. Feltrinelli, Milan
Chambers N, Simmons C, Wackernagel M (2000) Sharing nature's interest. Ecological footprints as an indicator of sustainability. Earthscan, London
Charter M, Tischner U (ed) (2001) Sustainable solutions: developing products and services for the future. Greenleaf, Sheffield
Cova B (1995) Au-delà du marché: quand le lien importe plus que le bien. L'Harmattan, Paris
Daly H, Cobb E (1989) For the common good. Beacon, Boston
De Leonardis O (1994) Approccio alle capacità fondamentali. In: Balbo L (ed) Friendly 94. Anabasi, Milan

Douglas M, Isherwood B (1979) The world of goods: towards and anthropology of consumption. Basic Books, New York
Durning A (1992) How much is enough? Earthscan, London
EMUDE, Emerging User Demands for Sustainable Solutions (2006) 6th Framework Programme (priority 3-NMP), European Community, final report
Florida R (2002) The rise of the creative class. And how it's transforming work, leisure, community and everyday life. Basic Books, New York
French H (2000) Vanishing borders. Protecting the planet in the age of globalization. Worldwatch Institute, Washington, DC
Fussler C, James P (1996) Driving eco innovation: a breakthrough discipline for innovation and sustainability. Pitman, London
Gershuny J (1978) After industrial society? The emerging self-service economy. MacMillan, London
Gesualdi F (1999) Manuale per un consumo responsabile. Feltrinelli, Milan
Giddens A (1991) The consequences of modernity. Polity, Cambridge
Giddens A (2000) Runaway world. How globalization is reshaping our lives. Profile, London
Hawken P (1993) The ecology of commerce. A declaration of sustainability. Harper Business, New York
Hawken P, Lovins A, Lovins H (1999) Natural capitalism. Creating the next industrial revolution. Little, Brown, Boston
Jansen JLA (CLTM) (1993) Toward a sustainable Oikos. En route with technology! Documento di Lavoro, Delft
Klein N (2002) No Logo. Picador, New York
Landry C (2000) The creative city. A toolkit for urban innovators. Earthscan, London
McCann-Ericksson/UNEP (2002) Can sustainability sell? UNEP Publications, Paris
McDonough W, Braungart M (2002) Cradle to cradle: remaking the way we make things. North Point, New York
McMichael T (2001) Human frontiers, environments and disease. Past patterns, uncertain futures. Cambridge University Press, Cambridge
Manzini E, Jegou F (2003) Sustainable everyday. Scenarios of urban life. Ed. Ambiente, Milan
Manzini E, Vezzoli C (2002) Product-service systems and sustainability. UNEP, Paris
Manzini E, Collina L, Evans E (eds) (2004) Solution oriented partnership, how to design industrialised sustainable solutions. Cranfield University, Cranfield
Marten GG (2001) Human ecology. Basic concepts for sustainable development. Earthscan, London
Matthews E, Hammond A (1999) Critical consumption trends and implications: degrading Earth's ecosystem. World Resources Institute, Washington, DC
Meroni A (2006) Creative communities. People inventing new ways of living. Ed. Polidesign
Mont O (2002) Functional thinking. The role of functional sales and product service systems for a functional based society, research report for the Swedish EPA, IIIEE Lund University, Lund
Moscovici S (1981) Psicologia delle minoranze attive. Bollati Boringhieri, Turin
Norman R, Ramirez R (1995) Le strategie interattive d'impresa. Etaslibri, Milan
Nussbaum MC, Sen A (eds) (1993) The quality of life. Clarendon, Oxford
Pauli G (1997) Breakthroughs – what business can offer society. Baldini, Castoldi, Milan
Pauli G (1999) Il Progetto Zeri. Il Sole 24 ORE, Milan
Perna T (1998) Fair trade. La sfida etica al mercato mondiale. Bollati Boringhieri, Turin
Pine JB, Gilmore JH (1999) The experience economy. Work is theatre and every business a stage. Harvard Business School Press, Boston
Princen T, et al. (eds) (2002) Confronting consumption. MIT Press, Cambridge
Ray PH, Anderson SR (2000) The cultural creatives, how 50 million people are changing the world. Three Rivers Press, New York
Rees W (1992) Ecological footprints and appropriate carrying capacities; what urban economics leaves out. Environ Urban 4:121–130

Rheingold H (1994) The virtual community. Sperling & Kupfer Editori, Milan
Rifkin J (2000) The age of access. Tarcher/Putnam, New York
Rifkin J (2002) The hydrogen economy. Putnam, New York
Sachs W (1992) The development dictionary. A guide to knowledge as power. Zed Books
Sachs W (1993) Global ecology. A new arena for the political conflict. Zed, London
Sachs W (1999) Planet dialectics. Explorations in environment and development. Zed, London
Sachs W (ed) (2002) The Jo'burg-Memo. Fairness in a fragile world. Memorandum for the World Summit on Sustainable Development. Heinrich Böll Foundation, Berlin
Sassen S (2002) Global networks, linked cities. Routledge, London
Schmidt-Bleek F (1993) MIPS re-visited. Fresenius Environ Bull 2
Schor J (1999) The overspent American. Upscaling, downshifting, and the new consumer. Harper, New York
Sen A (2004) Why we should preserve the spotted owl. London Review of Books 26(3)
Shiva V (1989) Staying alive: women, ecology and development. Zed, London
Shiva V (1993) Monocultures of the mind: perspectives on biodiversity and biotechnology. Zed, London
Solow RM (1992) An almost practical step towards sustainability, resources for the future. RFF, Washington, DC
Stahel RW (1997) The functional economy: cultural change and organizational change. In: Richards DJ (ed) The industrial green game. National Academic Press, Washington, DC
Sto E (ed) (1995) Sustainable consumption. SIFO, Lysaker
UNEP/Consumers International (2002) Tracking progress: implementing sustainable consumption policies. UNEP, Paris
Von Weizsacker E (1988) Come vivere con gli errori. Il valore evolutivo degli errori. In: Ceruti M, Laszlo E (eds) Physis: abitare la terra. Feltrinelli, Milan
Von Weizsacker E (1994) Earth Politics. Zed, London
Von Weizsacker EU, Lovins AB, Lovins LH (1997) Factor four: doubling wealth-halving resource use. Earthscan, London
Wackernagel M, Rees W (1996) Our ecological footprint. Reducing human impact on the earth. New Society, Gabriola Island
WBCSD (1993) Getting eco-efficient, report of the Business Council For Sustainable Development, First Antwerp Eco-Efficiency Workshop
WBCSD (1995) Eco-efficient leadership for improved economic and environmental performance. WBCSD, Geneva
WBCSD, Ayres RU (ed) (1995) Achieving eco-efficiency in business. Rapporto del World Business Centre for Sustainable Development Second Antwerp Eco-Efficiency Workshop, Anversa
WCED, World Commission on Environmental Development (1987) Our common future. Oxford University Press, Oxford
Wenger E, McDermott R, Snyder W (2002) Cultivating communities of practices. A guide to managing knowledge. Harvard Business School Press, Cambridge
Worldwatch Institute (2002) Vital signs 2002–2003. The trends that are shaping the future, Earthscan, London
Wuppertal Institut (1996) Zukunftsfähiges Deutschland. Birkhauser, Berlin-Basilea
Wuppertal Institut (1996) Per una civiltà capace di futuro. Birkhauser, Berlin-Basilea

Part II

AAVV (1992) Fare e disfare, conference proceedings. Politecnico di Milano, Milan
AAVV (1996) Architettura e natura. Progettare la sostenibilità. Ed. Nuove IniziaCtive, Milan

Andersen M (2006) System innovation versus innovation systems – challenges for the SCP agenda. Perspectives on radical changes to sustainable consumption and production (SCP). Sustainable Consumption Research Exchange (SCORE!) Network, Copenhagen

AAVV (1997) Refuse. Making the most of what we have. Cultural Connections, Utrecht

Alting L (1993) Life-cycle design of products: a new opportunity for manufacturing, Denmark, Danish Technical University, Life Cycle Centre, Institute for Product Development

Alting L, Boothroyd G (1993) Design for assembly and disassembly, chapter 1, Danish Technical University, Life Cycle Centre, Institute for Product Development

Alting L, Jorgensen J, Legarth J, Erichsen H, Gregersen J (1994) Development of environmental guidelines for electronic appliances, Denmark. Danish Technical University, Life Cycle Centre, Institute for Product Development, IEEE

Associazione impresa Politecnico (1995) Eco-efficienza ed innovazione, atti del convegno. Politecnico di Milano, Milan

Atakari I, Jackson AM, Vianco PT (1994) Evaluation of lead-free solder joints in electronic assemblies, USA. AT&T Bell Laboratories and Sandia national laboratories

Bistagnino L (2003) Design con un futuro. Time and Mind ed., Turin

Bistagnino L (2005) System design, un nuovo modello di business. Conference proceedings Formazione sviluppo sostenibile e design. Politecnico di Milano, Milan

Benjamin Y, Edirisinghe M, Zwetsloot G (1994) New material for environmental design, European Foundation for the Improvement of Living and Working Conditions, Dublin

Bertucelli L, van der Wal (1993) Tecnologie per il riciclo dei poliuretani. DOW Italia

Bianchi U (1993) Non oil = polimeri degradabili e non. In: La progettazione ambientalmente consapevole con le materie plastiche, Milan

Brezet H (2003) Product development with the environment as innovation strategy. The promise approach. Delft University of Technology, Delft

Brezet H, van Hemel C (1997) Ecodesign. A promising approach to sustainable production and consumption. UNEP, Paris

Brezet H, et al. (2001) The design of eco-efficient services. Methods, tools and review of the case study based "Designing Eco-efficient Services" project. VROM, The Hague

Brower C, Mallory R, Ohlman Z (2005) Experimental eco-design. Rotovision, Switzerland

Butera F (1979) Quale energia per quale società, le basi scientifiche per una politica energetica alternativa. Gabriele Mazzotta editore, Milan

Casati B (1997) Design, Plastica, Ambiente. Progettare per il ciclo di vita dei polimeri, Maggioli editore, Rimini

Catania A (2005) Eco-efficienza dei prodotti, conference proceedings. Formazione sviluppo sostenibile e Design, Politecnico di Milano

Charter M, Tischner U (2001) Sustainable solutions, developing products and services for the future, Greenleaf, Sheffield

Cipriani A (1994) Strategia omomateriale. Politecnico di Milano

Cooper T (1996) Beyond recycling. The longer life option. The New Economics Foundations, London

Cooper T, Sian E (2000) Products to Services, the friends of the earth. Centre for Sustainable Consumption, Sheffield Hallam University, Sheffield

Czekay J (1994) The life cycle concept as a potential strategy for a design process for sustainable products UIAH, Helsinki

Design World (1992) Plastic possibilities. Green design, (case study car and refrigerator). Design World no. 23

Di Carlo S, Serra R (1993) La progettazione ambientalmente consapevole con le materie plastiche: l'industria dell'auto. Fiat laboratori centrali, Turin

Dowie T (1993) Results of computer monitor disassembly, DDR/TR 4. Manchester Metropolitan University, Manchester

European Union (2003) Directive 2002/96/EC, WEEE, Waste Electrical and Electronic Equipment (RAEE in Italian). Official Journal of the European Union

European Union (2003) Directive 2002/95/EC, EC on the Restriction of the use of certain hazardous. substances in electrical and electronic equipment (RoHS). Official Journal of the European Union

European Union (2005) Directive 2005/32/EC, EUP re-establishing a framework for the setting of ecodesign requirements for energy-using products. Official Journal of the European Union

Feldmann K, Meedt O, Scheller H (1994) Life cycle engineering – challenge in the scope of technology. Economy and General regulation. In: Recy '94. FAPS, Erlangen

Fieschi M, Pretato U (2001) Politiche Integrate di Prodotto: un'impostazione per lo scenario italiano, ANPA, Rome

Fregola M, Orth P, Riess R (1995) Recupero funzionale dei tecnopolimeri: esempi applicativi. In: Plast

Fussler C (1995) Eco-efficiency a cure to marketing myopia. DOW Europe

Fussler C (1996) Driving eco innovation. A breakthrough discipline for innovation and sustainability, Pitman, London

Giudice F, La Rosa G, Risitano (2006) Product design for the environment. A life cycle approach. CRC/Taylor and Francis, Boca Raton

Hauschild M, Wenzel H, Alting L (1999) Life Cycle Design: a route to the sustainable industrial culture? CIRP Ann Manuf Technol 48(1):393–396

Heiskinen E (1996) Condition for product life extension, Working paper of the National Consumer Research Centre, Helsinki

Heiskinen EC (2002) The institutional logic of life cycle thinking. J Cleaner Prod 10(5):427–437

ISO (2002) 14062, Environmental management – integrating environmental aspects into product design and development, ISO/TR 14062:2002(E). ISO, Geneva

Joore P (2006) Guide me: translating a broad societal need into a concrete product service solution. In: Perspectives on radical changes to sustainable consumption and production (SCP). Sustainable Consumption Research Exchange (SCORE!) Network, Copenhagen

KW, et al. (1975) The efficient use of energy: a physics perspective. AIP conference proceedings, no. 25. American Institute of Physics, New York

Kahmeyer M, Leicht T (1992) Dismantling facilitated. Plast Europe 1992, Fraunhofer Institute for Production Technology and Automation, Newport

Kahmeyer M, Schmaus T (1992) Design for disassembly – challenge for the future. In: Forum on design for manufacturing and assembly. Fraunhofer Institute for Production Technology and Automation, Newport

Kahmeyer M, Warneke H-J (1993) Design For disassembly: a computer aided approach applied to computer hardware. In: 1993 Forum on design for manufacturing and assembly. Fraunhofer Institute for Production Technology and Automation, Newport

Kahmeyer M, Warneke H-J, Schweizer M (1992) Flexible disassembly with industrial robot. In: International Symposium on Industrial Robot, Barcelona-Ltd. Fraunhofer Institute for Production Technology and Automation, Newport

Kahmeyer M, Schraft RD, Degenhart E, Hägele M (1993) New robot applications in production and service. In: IIEEE/Tsukuba International Workshop on Advanced Robotics, Fraunhofer Institute for Production Technology and Automation, Newport

Karlsson R, Luttrop C (2006) EcoDesign: what is happening? An overview of the subject area of EcoDesign. J Cleaner Prod 14(15):1291–1298

Keoleian GA, Menerey D (1993) Life cycle design guidance manual. Environmental requirements and the product system. EPA, Washington, DC

Kuuva M (1993) Design for recycling. University of Technology, Helsinki

Kuuva M, Airila M (1994) Conceptual approach on design for practical product recycling, University of Technology, Helsinki

La Mantia P (1991) Riciclabilità di materie plastiche: degradazione ed incompatibilità. In: Riciclare 91: le Poliolefine

Langella C (2005) Design sostenibile per un sistema ecomuseale. Conference proceedings Formazione sviluppo sostenibile e design. Politecnico di Milano, Milan

Lanzavecchia C (2000) Il fare ecologico. Il prodotto industriale e i suoi requisiti ambientali, Paravia Bruno Mondatori Editore, Turin
La Rocca F (2005) Design e pluralismo tecnologico. Conference proceedings Formazione sviluppo sostenibile e design. Politecnico di Milano, Milan
Liberti R (2003) Design per la moda. Tecnologie e scenari innovativi. Alinea editore, Florence
Lindhqvist T (1993) L'estensione della responsabilità del produttore, atti del convegno Fare e disfare, Politecnico di Milano, Milan
Lotti G (ed) (2002) Progettare e produrre per la sostenibilità. Casa Toscana, Pisa
Mangiarotti R (2000) Design for environment. Maggioli editore, Rimini
Manzini E (1990) Artefatti – verso una nuova ecologia dell'ambiente artificiale. Domus Academy, Milan
Manzini E (1996) Ecodesign e minimizzazione dei rifiuti – la riprogettazione dei prodotti come soluzione strategica al problema dei rifiuti. Politecnico di Milano, Milano
Manzini E, Vezzoli C (1998) Lo sviluppo di prodotti sostenibili. I requisiti ambientali dei prodotti industriali. Maggioli editore, Rimini
Manzini E, Vezzoli C (2001) Product service systems as a strategic design approach to sustainability. Examples taken from the "Sustainable Innovation" Italian prize. Conference proceedings Towards Sustainable Product Design, Amsterdam
Manzini E, Collina L, Evans S (2004) Solution oriented partnership. How to design industrialised sustainable solutions. Cranfield University, Cranfield
Marano A (2005) Dal servizio al prodotto sostenibile. Conference proceedings Formazione sviluppo sostenibile e design. Politecnico di Milano, Milan
Marchand A, Walker S (2006) Designing alternatives. In: Perspectives on radical changes to sustainable consumption and production (SCP). Sustainable Consumption Research Exchange (SCORE!) Network, Copenhagen
Mayer C (1995) A case study of recycling and product life cycle, Mayer Cohen, UK
Ministere de l'Environment (1996) Catalogue de la prévention des déchets d'emballages, Paris
Molinari T (ed) (1997) Ri-usi. Triennale di Milano, Corriani Editore, Mantova
Mont O (2004) Product-service systems: panacea or myth? PhD thesis, University of Lund, Lund
Mont O, Emtairah T (2006) Systemic changes for sustainable consumption and production. In: Perspectives on radical changes to sustainable consumption and production (SCP). Sustainable Consumption Research Exchange (SCORE!) Network, Copenhagen
Normann R, Ramirez R (1995) Le strategie interattive d'impresa. Dalla catena alla costellazione del valore. Etas Libri, Milan
Nulli M (1993) Il riciclo meccanico dei termoplastici aspetti legati alle tecnologie di trasformazione. In: La progettazione ambientalmente consapevole con le materie plastiche. Politecnico di Milano, Milan
O2 (2006) Ecodesign gallery, O2 Global Network, http:\\www.ecomarkt.nl\europe\design
PIA (1991) Design for plastic recycling. PIA, Australia
Pietroni L (2004) Eco-materiali ed eco-prodotti "made in Italy". Ed. Kappa, Rome
Pietroni L (2005) Lo scenario dell'Ecodesign tra cultura, mercato e sperimentazione didattica. Conference proceedings Formazione sviluppo sostenibile e design. Politecnico di Milano, Milan
Pinetti L (1992) Il riciclaggio di materie plastiche in Italia. L'influenza dei fattori economici, tecnologici e legislativi sull'andamento del settore. Assorimap, Milan
Ranzo P (2005) L'alba di un nuovo primitivismo, conference proceedings Formazione sviluppo sostenibile e design. Politecnico di Milano, Milan
Rifkin J (2000) The age of access. How to shift from ownership to access in transforming capitalism. Penguin, London
Rocchi S (2005) Enhancing sustainable innovation by design. An approach to the co-creation of economic, social and environmental value. Thesis, Erasmus University, Rotterdam
Ryan C (2004) Digital eco-sense: sustainability and ICT – a new terrain for innovation. Lab 3000, Melbourne

Schmidt-Bleek F (1995) The role of industrial products on the way towards sustainability. Wuppertal Institut, Wuppertal
Schmidt-Bleek F, Hinterberger F, Kranedonk S, Welfens JM (1994) Increasing resources productivity through eco-efficient services, Wuppertal papers. Wuppertal Institut, Wuppertal
Scholl G (2006) Product service systems. Proceedings, perspectives on radical changes to sustainable consumption and production (SCP). Sustainable Consumption Research Exchange (SCORE!) Network, Copenhagen
Sekutowski JC (1994) Developing internal competence in DFE: a case study, USA. AT&T Bell Environmental Technology Department
Severini F, Coccia M (1991) Recupero postconsumo delle materie plastiche. Edizione IVR, Italy
Simon M, Dowie T (1992) Disassembly process planning, DDR/TR2. Manchester Metropolitan University, Manchester
Simon M, Dowie T (1993) Results of telephone disassembly, DDR/TR 9. Manchester Metropolitan University, Manchester
Simon M, Dowie T (1993) Quantitative assessment of design recyclability, DDR/TR 8. Manchester Metropolitan University, Manchester
Simon M, Dowie T (1993) Results of gas cooker disassembly, DDR/TR 5. Manchester Metropolitan University, Manchester
Simon M, Fogg B, Chambellant F (1992) Design for cost-effective disassembly, DDR/TR1. Manchester Metropolitan University, Manchester
Sondern AD (1993) Design for recycling, case study a TV. In: Styrenics '93, Philips Consumers Electronics
Stahel W (1991) Product innovation in the service economy. The Product Life Institute, Geneva
Stahel W (2001) Sustainability and services. In: Charter M, Tischner U (eds) Sustainable solutions. Developing products and services for the future. Greenleaf, Sheffield
Steinhilper R (1992) Recycling of high tech products. In: First NOH European Conference design for environment. Fraunhofer Institute for Production Technology and Automation, Newport
Summers CM (1971)
Surrey Institute of Art and Design. The Journal of Sustainable Product Design. The Centre for Sustainable Design, Surrey
Taskanen P, Takala R (2006) A decomposition of the end-of life process. J Cleaner Prod 14(15):1326–1332
Tischner U, Verkuijl M (2006) Design for (Social) Sustainability and Radical Change. In: Perspectives on radical changes to sustainable consumption and production (SCP). Sustainable Consumption Research Exchange (SCORE!) Network, Copenhagen
Tukker A, Tischner U (eds) New business for Old Europe. Product services, sustainability and competitiveness. Greenleaf, Sheffield
Uslenghi M (1993) Scelta del Polimero nell'ottica del recupero. In: Riciclo tecnopolimeri. Basf, Milan
Van Halen C, Vezzoli C, Wimmer R (eds) (2005) Methodology for product service system. How to develop clean, clever and competitive strategies in companies. Van Gorcum, Assen
Van Hemel CG (1998) Eco design empirically explored. Design for environment in Dutch small and medium sized enterprises. Delft University of Technology, Design for Sustainability Research Programme. Boekahandel MilieuBoek, Amsterdam
Van Nes N, Cramer J (2006) Product life time optimization: a challenging strategy towards more sustainable consumption patterns. J Cleaner Prod 14(15):1307–1318
Veneziano R (2005) High-low design e sostenibilità ambientale. Conference proceedings Formazione sviluppo sostenibile e design. Politecnico di Milano, Milan
Vezzoli C (1997) Life cycle design e informatica. In: Conferenza La progettazione di prodotti eco-efficiernti. SMAU, Milan
Vezzoli C (1997) Life cycle design, a strategy for the development of sustainable products: case study of a telephone, industry and environment review. United Nations Environmental Programme, Paris

Vezzoli C (1997) Environmental sustainability and product-service design strategies. International expert brainstorming seminar. New horizon in cleaner production. UNEP, Sweden
Vezzoli C (2002) From life cycle design to sustainable product-service system design. Approach and preliminary results of the EU research "MEPSS. A toolkit for industry". Conference proceedings International conference on ecodesign IEEP, New Delhi
Vezzoli C (2003) Designing systemic innovation for sustainability. In: Cumulus Working Papers, conferenza VALID, Value in design, Tallinn. Edito da University of Art and Design, Helsinki
Vezzoli C (2004) eco.DISCO. Il design per la sostenibilità ambientale. Agenzia Nazionale per la Protezione dell'Ambiente e servizi Tecnici, DIS-INDACO. Politecnico di Milano, Milan
Vezzoli C (2007)
Vezzoli C, Manzini E (2002) Product-service systems and sustainability. Opportunities for sustainable solutions. United Nations Environment Programme, Division of Technology Industry and Economics, Production and Consumption Branch, Paris
Vezzoli C, Manzini E (2006) Design for sustainable consumption. Conference proceedings of Prospettive di cambiamenti radicali per un consumo sostenibile. SCORE!, Copenhagen
Vezzoli C, Sciama D (2003) Domestico sostenibile. Abitare il tempo, Verona
Vezzoli C, Sciama D (2006) Life Cycle Design: from general methods to product type. Specific guidelines and checklists: a method adopted to develop a set of guidelines/checklist handbook for the eco-efficient design of NECTA vending machines. J Cleaner Prod 14(15):1319–1328
Von Weizsacker E, Lovins A, Lovins H (1997) Factor four. Doubling wealth – halving resource use. Earthscan, London
Wenzel H (1996) Life cycle diagnosis. Danish Technical University, Life Cycle Centre, Institute for Product Development, Kgs. Lyngby
Wimmer R, Kang M (2006) Product service systems as a holistic cure for obese consumption and production. In: Perspectives on radical changes to sustainable consumption and production (SCP). Sustainable Consumption Research Exchange (SCORE!) Network, Copenhagen
Zhang B, Simon M, Dowie T (1993) Initial investigation of design guidelines and assessment of design for recycling, DDR/TR 3. Manchester Metropolitan University, Manchester
Zwollinski P, Lopez-Ontiveros M-A, Brissaud D (2006) Integrated design of remanufacturable products based on products profiles. J Cleaner Prod 14(15):1333–1345

Part III

AAVV (1993) "MIPS". Fresenius Envir Bull 2(8)
Alting L, Wenzel H, Haushild MZ, Jorgensen J (1993) Environmental tool in product development, Danish Technical University, Life Cycle Centre, Institute for Product Development, 1993
Baldo G, Marino M, Rossi S (2005) Analisi del ciclo di vita. LCA, Ed. Ambiente
Benjamin Y (1995) Design for health profiler. European Foundation for Improvement of Living and Working Conditions, Dublin
Bianchi D (1995) Effetti ambientali della multimedialità, Milano per la Multimedialità. Quaderno no. 29, AIM, Milan
Boustead I (1995) Eco-profiles of the European plastics industry; rapporti 1–13. APME PWMI, Brussels
Boustead I, Consoli F, Allen D, de Oude N, Fava J, Franklin W, Quay B, Parrish R, Perriman R, Postlethwaite D, Seguin J, Vigon B (1993) Guidelines for life-cycle assessment: a code of practice, SETAC-Europe, Brussels
Bright C, Flavin C, Platt McGinn A, et al. (2003) State of the world 2003. A Worldwatch Institute report on progress toward a sustainable society. Worldwatch Institute, Washington, DC

Bringezu S (1993) Resources intensity analysis. A screening step for LCA. Wuppertal Institut, Wuppertal
Buwal (1991) Ecobalance of packing materials state of 1990, n° 132. Buwal, Bern
Buwal (1991) Methodologie des ecobilans sur la bas de l'optimisation écologique, no. 133, Buwal, Bern
CML, NOVEM, RIVM (1996) Life Cycle Assessment: what is and how to do it. United Nation Environmental Programme, Industry and the Environment, Paris
Dahlström H (1999) Company-specific guidelines. J Sustain Prod Design 8:18–24
Donnelly K, Beckett-Furnell Z, Traeger S, Okrasinski T, Holman S (2006) Ecodesign implemented through a product-based environmental management system. 14(15–16):1357–1367
EPA (1993) Life-Cycle Assessment: inventory guidelines and principles, EPA/600/R-92/245. EPA, Cincinnati
European Union (2005) Directive 2005/32/CE, EUP relativa all'istituzione di un quadro per l'elaborazione di specifiche per la progettazione ecocompatibile dei prodotti che consumano energia. Official Journal of the European Union
FEB (1993) MIPS. Fresenius Envir Bull 2(8)
Giudice F, La Rosa G, Risitano A (2006) Product design for the environment. A life cycle approach. CRC/Taylor and Francis, Boca Raton
Guinée JB, Heijungs R, Udo de Haes H, Huppers G (1993) Quantitative life cycle assessment of products 1. Goal definition and inventory. Leiden University, Leiden
Guinée JB, Heijungs R, Udo de Haes H, Huppers G (1993) Quantitative life cycle assessment of products 2. Classification, valuation and improvement analysis. Leiden University, Leiden
Guinée JB Impact assessment within the framework of life cycle assessment of products. In: Seminar & Workshop Energy Issues in Life Cycle Assessment. Leiden University, Leiden
ISO 14040, Environmental management – Life Cycle Assessment – principles and framework. ISO/TR 14040:1997(E). ISO, Geneva
istituto Italiano Imballaggio (1995) E' nato GreenPack, in presentazione GreenPack
Johansson G, Magnusson T (2006) Organising for environmental considerations in complex product development projects: implications for introducing a green sub-project. J Cleaner Prod 14(15):1368–1376
Kahmeyer M, Warneke H-J (1993) Design for disassembly: a computer aided approach applied to computer hardware. In: 1993 Forum on Design for Manufacturing and Assembly. Fraunhofer Institute for Production Technology and Automation, Newport
Karlsson R, Luttrop C (2006) EcoDesign: what is happening? An overview of the subject area of eco design. J Cleaner Prod 14(15):1291–1298
Lofthouse V (2006) Ecodesign tools for designers: defining the requirements. J Cleaner Prod 14(15):1386–1395
Manzini E, Vezzoli C (1998) Lo sviluppo di prodotti sostenibili. I requisiti ambientali dei prodotti industriali. Maggioli editore, Rimini
Manzini E, Collina L, Evans S (2004) Solution oriented partnership. How to design industrialised sustainable solutions. Cranfield University, Cranfield
NOVEM, RIVM (1996) The ecoindicator 95. National Reuse of Waste Research Programme, Netherlands
SETAC (1992) Life-Cycle Assessment. SETAC, Brussels
SETAC (1993) Guidelines for Life-Cycle Assessment: a "code of practice". SETAC, Brussels
SPOLD (1996) The LCA sourcebook. SPOLD, Brussels
Sun J, et al. (2003) Design for environment: methodologies, tools, and implementation. J Integrated Design Process Science 7(1):59–75
Tingstrom J, Swanstrom L, Karlsson R (2006) A sustainability management in product development projects. The ABB experience. J Cleaner Prod 14(15):1377–1385
Tischner U, Schmidt-Bleek F (1993) Designing goods with MIPS. Wuppertal Institut, Wuppertal
United Nations Division for Sustainable Development (2001) Indicators of sustainable development: guidelines and methodologies. UN, New York

Van Halen C, Vezzoli C, Wimmer R (eds) (2005) Methodology for product service system. How to develop clean, clever and competitive strategies in companies. Van Gorcum, Assen
Van Hemel CG (1998) Eco design empirically explored. Design for environment in Dutch small and medium sized enterprises. Delft University of Technology, Design for Sustainability Research Programme, Boekahandel MilieuBoek Amsterdam
Vezzoli C (1997) Life Cycle Design e informatica, atti dalla conferenza La progettazione di prodotti eco-efficiernti. SMAU, Milan
Vezzoli C, Proserpio C (2004) Luce e Impatto ambientale. Una nuova generazione di artefatti per una luce sostenibile, Luce e design, no. 2. tecniche nuove editore, Milan
Vezzoli C, Proserpio P (2006) Il design degli elettrodomestici per la sostenibilità. In: Rampino L (ed) Sapere, Immaginare, Fare. Il Design d'innovazione per l'elettrodomestico. Ed. Polidesign, Milan
Vezzoli C, Sciama D (2003) Domestico sostenibile. Abitare il tempo, Verona
Vezzoli C, Sciama D (2006) Life Cycle Design: from general methods to product type specific guidelines and checklists: a method adopted to develop a set of guidelines/checklist handbook for the eco-efficient design of NECTA vending machines. J Cleaner Prod 14(15):1319–1325
Zhang B, Simon M, Dowie T (1993) Disassembly & assembly sequence generation by LPM, DDR/TR 6. Manchester Metropolitan University, Manchester

Part IV

AAVV (2002) Fare e disfare. Politecnico di Milano, Milan
Bistagnino L (2004) Design con un futuro. Time and Mind, Turin
Bistagnino L (2005) System design, un nuovo modello di business. Conference proceedings Formazione sviluppo sostenibile e design. Politecnico di Milano, Milan
Bistagnino B, Lanzavecchia C (2000)
Bresso M (1993) Per un'economia ecologica, NIS
Brezet H, van Hemel C (1997) Ecodesign. A promising approach to sustainable production and consumption. UNEP, Paris
Brezet H, et al. (2001) The design of eco-efficient services. Methods, tools and review of the case study based "Designing eco-efficient Services" project. VROM, The Hague
Butrera M (2005) Fisica, tecnica e design. Conference proceedings Formazione sviluppo sostenibile e design. Politecnico di Milano, Milan
Catania A (2005) Eco-efficienza dei prodotti. Conference proceedings Formazione sviluppo sostenibile e design. Politecnico di Milano, Milan
Charkiewicz E (2002) Changing consumption patterns in Central and Eastern Europe. UNEP, Industry and Environment Review, no. January–March
Charter M, Tischner U (2001) Sustainable solutions. Developing product and services for the future. Greenleaf, Sheffield
Cooper T, Sian E (2000) Products to services, the friends of the earth. Centre for Sustainable Consumption, Sheffield Hallam University
Costa F, Vezzoli C (2001) Formazione e domanda di professionalità ambientali nel settore del disegno industriale. Relazione sullo stato dell'arte dell'insegnamento della disciplina dei Requisiti ambientali dei prodotti industriali e della presenza di competenze progettuali-ambientali nelle aziende e negli studi di design. CIR.IS-INDACO Politecnico di Milano, ANPA, Milan, scaricabile da
www.polimi.it/rapirete sezione relazioni
Dahlström H (1999) Company-specific guidelines. J Sustain Prod Design 8:18–24
ERL (1994) Every decision counts: consumer choice and environmental impacts. Ministerie van Volkshuisvesting, Oxford

ERL (1994) The best of both worlds: sustainability and quality. Lifestyles in the 21st century. Oxford, 1994
European Union (2003) Directive 2002/95/EC, the restriction of the use of certain hazardous. substances in electrical and electronic equipment (RoHS). Official Journal of the European Union
European Union (2003) Directive 2002/96/EC, WEEE, Waste electrical and electronic equipment, Official Journal of the European Union
European Union (2005) Directive 2005/32/EC, EUP re-establishing a framework for the setting of ecodesign requirements for energy-using products. Official Journal of the European Union
Florida R (2002) The rise of the creative class. And how it is transforming work, leisure, community and everyday life. Basic Books, New York
Geyer-Allely E (2002) Sustainable consumption: an insurmountable challenge. UNEP, Industry and Environment Review, no. January–March
Giudice F, La Rosa G, Risitano A (2006) Product design for the environment. A life cycle approach, CRC/Taylor and Francis, Boca Raton
Hawken P (1995) The ecology of commerce. A declaration of sustainability. Phoenix, London
Keoleian GA, Menerey D (1993) Life cycle design guidance manual. Environmental requirements and the product system. Environmental Protection Agency, Washington, DC
Langella C (2005) Design sostenibile per un sistema ecomuseale, conference proceedings Formazione sviluppo sostenibile e Design. Politecnico di Milano, Milan
Lanzavecchia C (2000) Il fare ecologico. Il prodotto industriale e i suoi requisiti ambientali. Paravia Bruno Mondatori Editore, Turin
La Rocca F (2005) Design e pluralismo tecnologico, conference proceedings Formazione sviluppo sostenibile e design. Politecnico di Milano, Milan
La Rocca F (2006) Il tempo opaco degli oggetti. Forme evolutive del design contemporaneo. Franco Angeli, Milan
Liberti R (2005) Proposte di didattica/ricerca sperimentale per una progettazione sostenibile. Conference proceedings Formazione sviluppo sostenibile e design. Politecnico di Milano, Milan
Liberti R (2003) Design per la moda. Tecnologie e scenari innovativi. Alinea editore, Florence
Lotti G (ed) (2002) Progettare e produrre per la sostenibilità. Casa Toscana, Pisa
Lotti G (2005) Design per la sostenibilità, qui Firenze. Conference proceedings Formazione sviluppo sostenibile e design. Politecnico di Milano, Milan
Maldonado T (1970) La speranza progettuale. Ambiente e società. Einaudi, Turin
Maldonado T (1998) Disegno Industriale, Questione ambientale e didattica, in atti seminario Design, didattica e questione ambientale. Agenzia Nazionale per la Protezione dell'Ambiente, Politecnico di Milano, Milan
Mangiarotti R (2000) Design for environment. Maggioli editore, Rimini
Manzini E (2003) La società creativa. Il mondo come potrebbe essere, atti del seminario Innovazione sociale sostenibile. Iniziative e progetti a confronto tra realtà milanesi e università, Unità di Ricerca DIS e RAPI.labo Politecnico di Milano, Forum CONSUMOCRITICO e Insieme nelle Terre di Mezzo, Milan
Manzini E (2005) Design sostenibilità e innovazione sociale. Conference proceedings Formazione sviluppo sostenibile e design. Politecnico di Milano, Milan
Manzini E, Jegou F (2003) Sustainable everyday. Scenarios of urban life. Ed. Ambiente, Milan
Manzini E, Vezzoli C (1998) Lo sviluppo di prodotti sostenibili. I requisiti ambientali dei prodotti industriali. Maggioli editore, Rimini
Manzini E, Vezzoli C (2001) Product Service Systems as a strategic design approach to sustainability. Examples taken from the "Sustainable Innovation" Italian prize. Conference proceedings Towards Sustainable Product Design, Amsterdam
Manzini E, Collina L, Evans S (2004) Solution oriented partnership. How to design industrialised sustainable solutions. Cranfield University, Cranfield
Marano A (ed) (2004) Design e ambiente. La valorizzazione del territorio tra storia umana e natura. Ed. Polidesign, Milan

Marano A (2005) Dal Servizio al prodotto sostenibile. Conference proceedings Formazione sviluppo sostenibile e design. Politecnico di Milano, Milan
Mont O (2004) Product-service systems: panacea or myth? PhD thesis, University of Lund, Lund
Normann R, Ramirez R (1995) Le strategie interattive d'impresa. Dalla catena alla costellazione del valore. Etas Libri, Milan
Papanek V (1973) Progettare per il mondo reale. Il design come è e come potrebbe essere. Mondadori, Milan
Penin L (2006) Strategic design for social sustainability in emerging contexts. PhD thesis, Politecnico di Milano, Milan
Pietroni L (ed) (2003) Eco e Bio Packaging. Quando il design incontra il cartone. COMIECO, Milan
Pietroni L (2004) Eco-materiali ed eco-prodotti "made in Italy". Ed. Kappa, Rome
Pietroni L (2005) Lo scenario dell'Ecodesign tra cultura, mercato e sperimentazione didattica. Conference proceedings Formazione sviluppo sostenibile e design. Politecnico di Milano, Milan
Pietroni and Vezzoli (2004)
Polidori C (2003) Il design qualunque. Ed. Union Printing, Rome
Proserpio C (2005) RAPI.rete e LENS, dalle reti nazionali alle reti internazionali. Conference proceedings Formazione sviluppo sostenibile e design. Politecnico di Milano, Milan
Ranzo P (2005) L'alba di un nuovo primitivismo. Conference proceedings Formazione sviluppo sostenibile e design. Politecnico di Milano, Milan
Rifkin J (2000) The age of access, how to shift from ownership to access in transforming capitalism. Penguin, London
Rifkin J (2002) The hydrogen economy. Putnam, New York
Rocchi S (2005) Enhancing sustainable innovation by design. An approach to the co-creation of economic, social and environmental value. Doctoral thesis. Erasmus University, Rotterdam
Ryan C (2004) Digital eco-sense: sustainability and ICT – a new terrain for innovation. Lab 3000, Melbourne
Sachs W, et al. (2002) The Jo'burg-Memo. Fairness in a Fragile World. Memorandum for the World Summit on Sustainable Development. Heinrich Böll Foundation, Berlin, www.boell.de
Stahel W (2001) Sustainability and services. In: Charter M, Tischner U (eds) Sustainable solutions – developing products and services for the future. Greenleaf, Sheffield
Tamborrini P (2005) L'impegno dell'ecodesign per una società sostenibile. Conference proceedings Formazione sviluppo sostenibile e design, Politecnico di Milano, Milan
Teatino A (2002) SPL ed LCD. Competenze, strumenti e strategie progettuali per la diffusione del Life Cycle Design nei Sistemi produttivi Locali, dottorato di ricerca in Disegno Industriale e Comunicazione Multimediale XIV° Ciclo, Facoltà del Design, Politecnico di Milano, Milan
Tukker A, Eder P (2000) Eco-design: Europena state of the art, ECSC-EEC-EAEC, Brussels
UN (2002) World summit on sustainable development. Draft political declaration. UN, Johannesburg
UNEP (2000) Achieving sustainable consumption patterns: the role of the industry. UNEP IE/IAC, Paris
Van Halen C, Vezzoli C, Wimmer R (eds) (2005) Methodology for product service system. How to develop clean, clever and competitive strategies in companies. Van Gorcum, Assen
Veneziano R (2005) High-low design e sostenibilità ambientale. Conference proceedings Formazione sviluppo sostenibile e design. Politecnico di Milano, Milan
Vezzoli C (2003) Designing systemic innovation for sustainability. In: Cumulus Working Papers, conference VALID, Value in design, Tallinn. University of Art and Design, Helsinki
Vezzoli C (2003) A new generation of designer: perspective for education and training in the field of sustainable design. Experiences and projects at the Politecnico di Milano University. J Cleaner Prod 11(1):1–9

Vezzoli C (2003) Design e sostenibilità. In: Manzini E, Bertola P (eds) Design multiverso. Appunti di fenomenologia del design. Ed. Polidesign, Milan

Vezzoli C (2003) Bello sostenibile! Ipotesi per il contributo dell'università del design all'innovazione sociale sostenibile, atti del seminario Innovazione sociale sostenibile. Iniziative e progetti a confronto tra realtà milanesi e università. Unità di Ricerca DIS e RAPI.labo Politecnico di Milano, Forum CONSUMOCRITICO e Insieme nelle Terre di Mezzo, Milan

Vezzoli C (ed) (2004) eco.DISCO. Il design per la sostenibilità ambientale. Agenzia per la Protezione dell'Ambiente e per i servizi Tecnici, DIS-INDACO Politecnico di Milano, ed. Poli.design, Milan

Vezzoli C (2005) Design di sistema per la sostenibilità: possibili integrazioni tra dimensione socioetica e ambientale. In: Design social. Universitade Estatual de Londrina, http://www.uel.br/eventos/workdesign/

Vezzoli C (2005) A migrant design creative-thought. Fostering necessary discontinuity for sustainability. Proceedings of Research in and through the arts. ELIA and Universität der Künste, Berlin

Vezzoli C (2007)

Vezzoli C, Manzini E (2002) Product-service systems and sustainability. Opportunities for sustainable solutions. United Nations Environment Programme, Division of Technology Industry and Economics, Production and Consumption Branch, Paris

Vezzoli C, Manzini E (2006) Design for sustainable consumption. New roles designing system innovations. In: Perspectives on radical changes to sustainable consumption and production (SCP). Sustainable Consumption Research Exchange (SCORE!) Network, Copenhagen

Vezzoli C, Penin L (2004) Campus: "lab" and "window" for sustainable design research and education. The DECOS educational network experience. In: Conference proceedings of EMSU2004. Environmental Management for Sustainable Education, Monterrey

Vezzoli C, Penin L (eds) (2005) Designing sustainable product-service systems for all. Libreria-Clup ed., Milan, clup@galactica.it

Vezzoli C, Soumitri GV (2002) Sustainable product-service system design for all. A design research working hypothesis. In: International conference on Ecodesign IEEP, New Delhi

Index

D

E

F

G

H

I

K

L

M

N